Agronomic Research for Food

ASA Special Publication Number 26

Papers presented at the annual meeting of the American Society of Agronomy in Knoxville, Tennessee, August, 1975

Editor: **Fred L. Patterson**
Editor-in-Chief: **Matthias Stelly**
Managing Editor: **David M. Kral**
Assistant Editor: **Linda C. Eisele**

1976

Published by the

AMERICAN SOCIETY OF AGRONOMY
677 South Segoe Road
Madison, Wisconsin 53711

Library of Congress Catalog Card Number: 76-9269
Standard Book Number: 0-89118-044-3

Contents

Foreword

The continuing food scarcity and the predicted food crisis in years ahead prompted the theme, "Agronomic Research for Food," for the 1975 annual meetings of the American Society of Agronomy, Crop Science Society America, and Soil Science Society of America.

Nearly 1,000 scientific papers were presented at the Knoxville meetings. There were symposia, joint sessions of divisions, and technical discussions that brought to bear the special expertise of 145 invited speakers from 14 countries.

Specific objectives were to provide a forum for in-depth review of fundamental discoveries in crops, soils, climate, and environment as well as present new ideas in production, management, and resources for food. New and imaginative approaches in student and adult education were discussed. Efforts were made through invited speakers to make the forums truly international in scope. The stimulation of formal and informal discussions between scientists, graduate students, and undergraduate students can lead to increased vigor and dedication for the 3,000 agronomists and guests who attended.

This special publication features the "overviews" of four invited speakers on "Agronomic Research for Food." Assuring an adequate food supply is one of the basic responsibilities of government. Some of the most pressing problems in expanding world food supply occur in the developing countries. It is here that agronomic advances, technology, resource allocation, and education are critical. Dr. Schuh discusses "Planning for Hunger Prevention" by appraising the current world agricultural situation, evaluating the changed conditions in the U. S., and detailing the steps needed to deal with world food needs.

The need for more efficient use of scarce agricultural inputs is clear. Dr. Barber gives an overview on "Efficient Fertilizer Use." He appraises the world fertilizer resources, the efficiency of fertilizer uptake, and research opportunities for the modification of soil and plants for more efficient fertilizer use.

The total water supply on this planet is essentially fixed, and interest in water management is worldwide. Dr. Hagan discusses the complexity of water management, the difficult public decisions on allocation of water to agriculture versus other uses, and opportunities for greater efficiency of water use in agriculture.

Scientists worldwide are alarmed at the rate of decrease in diversity of crop plants. Dr. Reitz discusses problems and programs for improving our scarce and diminishing plant germplasm resources for improving and protecting our food supply.

The majority of the 1,000 papers presented at the 1975 annual meeting will be published in the scientific journals or special publications of the three Societies. The quality and scope of these papers demonstrate the eagerness and dedication with which agronomists accept the challenge of Agronomic Research for Food.

Fred L. Patterson, President
American Society of Agronomy

Planning for Hunger Prevention

G. Edward Schuh

The problem of excess production at politically acceptable prices for agricultural products has plagued this country throughout most of the last 25 years. The legacy of this chronic problem is still with us today. It wasn't but a few short months ago that the President vetoed an Emergency Farm Bill designed to raise target and loan prices for agricultural products—legislation that was premised on the assumption that U. S. farm prices were going to drop to disastrously low levels unless the government intervened.

Since the first of July, 1975, farm prices have moved upward in an almost uninterrupted pattern. Spot prices for wheat (*Triticum aestivum* L.) have increased 40%, the price of corn (*Zea mays* L.) has increased 15%, and the price of soybeans (*Glycine max* L. Merr.) has increased 25%. Of the major commodities, only the price of beef is at a lower level today than it was in early July. Moreover, a hold has been put on further sales of grain to the Soviet Union until we have a firmer estimate of our own agricultural output.

Our present predicament is not unique. In 1973, the Secretary of Agriculture was reluctant to release all set-aside land back to production on the grounds that we would have excess production if he did so. Although all the land was released to production in 1974, there were similar concerns early in that year about potentially low prices for farmers.

You all know the record. We had to put a short-lived embargo on soybeans in the second half of 1973 because price increases were so great. Despite a record world output of grain in the crop year ending in 1973, farm prices did not decline to disastrously low levels. And in 1974, we had our own weather-induced shortfall, with the result that we again imposed export controls. The system of controls was put in place after we embargoed further exports to the Soviet Union in October of 1974.

The title your President-elect asked me to speak to is therefore more than appropriate: Planning for Hunger Prevention. Despite the progress we have made at home in providing food to all income groups, and the admirable performance of world agriculture throughout the post-World War II period, it is clear that there is a fine line between adequate food production for all, and a disaster of heretofore unimagined proportions.

G. Edward Schuh is senior staff economist, President's Council of Economic Advisers, on leave from Purdue University.

The world has been surprisingly free of major famines in this century. And many of the famines that did occur could have been prevented or largely alleviated if the world had known of their existence in time, or if there had been reasonably adequate means of transportation to the locale.

Similarly, per capita food production has increased at a modest rate of 0.4% in the developing or low-income countries for the 20-year period 1952 through 1972—despite a rate of population growth that is unprecedented in world history (Economic Research Service, 1974; Johnson, 1975).[1] Equally as important, there has been a dramatic increase in life expectancy in the developing countries. From about 1950 through 1970–75, life expectancy at birth increased by 40% in the low-income countries, with most of that increase due to a decline in death rates among the young.

On the other side of the coin, however, the U. S. Department of Agriculture (USDA) estimates that about one out of six people in the world is undernourished. And in the Assessment for the World Food Conference it was estimated that approximately 25% of the population in the developing market economies have less than adequate protein and energy consumption (United Nations, 1975).

Similarly, end-of-year stocks of total grain around the world have been estimated at only slightly over 100 million metric tons as we came out of the 1974–75 crop year—in marked contrast to the 176 million metric tons that we carried out of the 1960–61 crop year (Foreign Agricultural Service, 1975). Though there may be some slight rebuilding this year, the prognosis is not good. Hence, for the third year in a row we will end up at near pipeline levels. That means we will continue to be quite vulnerable to production shortfalls in any of the major producing or consuming countries.

In the rest of this paper I would like to divide my comments into three parts. First, I would like to make some comments on the world agricultural situation. That will be followed by a discussion of the changed conditions in U. S. agriculture. Finally, I would like to address specifically some of the things we need to do to deal with the problems we face.

THE WORLD AGRICULTURAL SITUATION

It is not sufficiently well recognized that food production has been expanding over the last two decades at exactly the same rate in the developing or low-income countries as in the developed countries. In the period from 1952 to 1962, the growth rate in both sets of countries was 3.1%/year. And in the period from 1962 to 1972, it was 2.7% in both sets of countries.

The difference between the two groups of countries is in their respective population growth rates. The developed countries had a population growth rate of 1.3% in the earlier period, with the result that per capita production increased at a rate of 1.8%. In the developing countries, population was in-

[1]Unless otherwise noted data in this paper have been taken from the two surveys of the world food problem cited.

creasing at a rate of 2.4%/year—almost double that of the advanced countries —with the result that per capita production increased at a rate of only 0.7%, only slightly over a third as great as for the advanced countries.

The consequence of the same decline in growth rates in food output between the 1950's and the 1960's was quite different for the two groups of countries. For the advanced countries, the growth rate of population declined by 30%, from 1.3 to 1.0%/year. As a result, per capita food production declined by only 0.1%, from 1.8 to 1.7%. In the developing or low-income countries, on the other hand, population growth rates remained the same. Consequently, the growth rate in per capita production declined by almost 60%, from 0.7 to 0.3%. The growth rate in per capita production clearly widened between the two groups of countries from the earlier decade to the more recent decade.

There are other important differences between the two sets of countries. For example, output in the low-income countries has been expanded largely by bringing additional land into production, and by increasing the agricultural labor force. Increases in productivity have played a much lesser role. In the advanced countries, however, land under cultivation has tended to remain static or actually decline, while the agricultural labor force has been reduced, in some cases dramatically. In these countries, the bulk of the increases in output have come from increases in productivity.

A simple comparison of grain yields between the two sets of countries tells most of the story. In the period 1934–38, grain yields per hectare were the same in the low-income and advanced countries, approximately 1.15 metric tons. By 1952–56, grain yields in the industrial countries had increased to 1.37 metric tons/ha, or almost 20%. Yields remained static in the low-income countries. By 1969–70, grain yields were up to 2.14 metric tons/ha in the advanced countries, an increase of another 86%. In the developing countries, however, yields by this date had increased to only 1.41 metric tons/ha. This constituted an increase of 22%. Although yields finally started to move up in less developed countries, their rate of increase was still approximately only one-fourth the rate of increase in the advanced countries.

This dimension of the differential performance in agriculture among the groups of countries constitutes the major challenge we face on the world food scene. A large part of the world's population, close to 2 billion people, is concentrated in the rice (*Oryza sativa*) producing and consuming centers of the world. This area includes Bangladesh, India, Indonesia, Pakistan, and the People's Republic of China. With the exception of Indonesia, every one of these countries has rather serious land constraints. Further output expansion has to come primarily by increasing productivity. Although there is a great deal of land that could be brought into production in other parts of the world, it isn't located where the people are. And trade is not a viable alternative unless the wherewithall to purchase the imports can be acquired.

Finally, a third difference between the advanced and the low-income or developing countries is in the economic policy with respect to agriculture. The advanced countries tend to protect and/or to subsidize their agricultural sectors, while the low-income countries tend to discriminate against theirs, both by failing to make the appropriate developmental investments and by discrimination through trade and price policy. This difference is worth keeping in mind when we turn to policy measures that can help us out of our present problems.

THE CHANGED CONDITIONS IN U. S. AGRICULTURE[2]

Agriculture has always been one of the more dynamic sectors of the U. S. economy. But in recent years there have been some fundamental changes not only within agriculture itself, but in how the agricultural sector relates to the larger national economy and to the world economy at large. To discuss all of these changes, even in a brief way, would require a rather long paper. But for our purposes here it is important that we discuss the decline in excess capacity in U. S. agriculture, one of the major developments of the last decade.

As was noted earlier, U. S. agriculture has been plagued by excess production at prevailing price ratios throughout the post-World War II period. The government was required to acquire large stocks and to remove them from the market in order to sustain prices at acceptable levels. Land was taken out of production and put in set-aside in an attempt to bring supply into balance with demand. And we subsidized exports and sold large quantities of our output on concessional terms just to be rid of it. A lot more we just gave away.

The reasons for this excess production are now reasonably well understood. In the first place, the nature of economic development is such that as an economy develops, labor has to be transferred out of agriculture, especially if productivity is growing in that sector. Unless labor can be transferred at a sufficiently rapid rate, incomes in agriculture will decline. In a real sense, it was the inability to transfer labor at a sufficiently rapid rate that caused supply to outrun demand at politically acceptable prices.

The problem was compounded because the U. S. has traditionally made sizable investments in research and development in agriculture. Although returning a high social rate of return to society (Arndt & Ruttan, 1975), these investments compounded the labor adjustment problem. When combined with a tendency to underinvest in the schooling of rural people, the result was a heavily burdened farm-nonfarm labor market. A side effect was that the agricultural labor force was forced to bear a major share of the adjustment costs of technical change—a technical change that was of great value to the economy as a whole.

Finally, agriculture suffered serious hidden discrimination throughout most of the post-World War II period by means of the overvaluation of the dollar (Schuh, 1974). The overvalued dollar caused all of our exports to be

[2]Material from this section is taken from G. Edward Schuh, 1975. The role of agricultural, food, and research policy in meeting food needs. Presented at the Working Conference on Research to Meet U. S. and World Food Needs, Kansas City, Missouri, July 9–11. More detail can be found in that paper, and in Chapter 6 of the *Economic Report of the President* (U. S. Government Printing Office, 1975).

less competitive in world markets than they otherwise would have been. Since agriculture was so dependent on export markets, and was already burdened with a serious labor adjustment problem, it experienced serious problems as a result of chronic disequilibrium in international money markets.

The paradox, of course, is that most people believe that agriculture and farm people have been major recipients of the largesse of the federal budget. In point of fact, however, the price support programs probably did little more than offset the discriminatory effect of the overvalued exchange rate. The same overvaluation of the dollar caused an important share of the benefits of technical change to be transferred to the U. S. consumer, rather than to the producer, where they would have naturally accrued. And for the most part society underinvested in such things as rural education, rural health, and other social services.

The devaluation of the dollar removed at least part of this discrimination. Our agricultural products are now much more competitive in world markets, with the result that now our problem is whether we are willing to export all that the foreign markets are willing to take. In each of the last 2 years we have imposed temporary export controls, and we may be forced to do it again this year.

It is important to note that at the same time that we experienced this once-for-all increase in the demand for farm output, a number of factors have affected the supply side of agriculture which suggest that output may not be increasing as rapidly in the future as it has in the past. In the first place, we discovered that much of what we believed was a 24.3 million hectare (60 million acre) reserve of land was in fact illusory. When this land was released to production, only 15 million hectares (37 million acres) actually come back into production.

Similarly, the excess labor has largely been drained out of U. S. agriculture. Average per capita incomes in the agricultural sector were larger than those in the nonfarm sector in both 1973 and 1974, probably for the first time in modern history. Median family incomes of farm people, while still less than median incomes of nonfarm families, have risen dramatically relative to those in the nonfarm sector.[3] The rate of outmigration from agriculture has slowed dramatically since 1970 (U. S. Government Printing Office, 1975). These are the criteria that economists use to determine whether a labor market has reached equilibrium.

The current high levels of unemployment in the general economy will undoubtedly temporarily disrupt this long-sought equilibrium in the agricultural labor market, since it will cause labor to be dammed up in agriculture again. But as the economy recovers, the outmigration will increase again, and when it does we will find that a major source of past slack in the agricultural sector has disappeared.

Finally, there has been a marked and little understood decline in the rate of measured productivity growth in U. S. agriculture. During the decade

[3]From 1970 to 1973, median family income of the farm population increased by 30% in real terms, compared to 6% in the nonfarm sector.

of the 1950's, total factor productivity grew by 27%. In the decade of the 1960's, it grew by only 11%, or at a rate only slightly more than a third as large as in the previous decade.

Increases in total factor productivity have been an important source of our output expansion in the past. From 1940 to 1970, for example, total measured physical inputs in agriculture increased only 4%, while output increased 58%. If the observed decline in productivity growth is real, and if it persists into the future, output increases in the future will depend more on increased use of conventional resources. This in turn will put agriculture in more direct competition with the rest of the economy in bidding for resources than it has been in the past.

The combination of an illusory land reserve, an emerging equilibrium in the labor market, and a decline in our rate of productivity growth suggests that the aggregate supply of agricultural output will increase less rapidly in the future than it has in the past. It should be noted, moreover, that it is the simultaneous *convergence* of these three factors that is important. Any one of them alone could probably be accommodated with relative ease. But occurring concurrently as they do poses major challenges to both economic and research policy.

ELEMENTS OF A POLICY FOR THE FUTURE

Despite this decline in the excess capacity of U. S. agriculture, we are in no danger of facing a hunger problem at home. We currently export the output from one out of every three hectares of farmland, and one out of every four dollars of farm income is earned from foreign markets. Clearly, exports could be cut back as our population grows—if that should be needed—and there would be a sizable cushion from this source. Moreover, the U. S. consumer still spends the smallest fraction of his budget on food of any country in the world, although this share has been increasing in recent years.

These conditions should not make up sanguine about our future, however. As long as we have a floating exchange rate and maintain free trade in agricultural products, we will share in the problem of world hunger whether we like it or not. Although our own food stamp and other income protection programs will probably keep the incidence of hunger from increasing here at home, we will still be subject to the consequences of shortfalls in output in other parts of the world, and be forced to pay a higher fraction of our budget for food when agricultural prices rise. Hence, it is important that we take an international perspective to our problems.

U. S. exports make a significant contribution to feeding the people in other lands. The now relatively small amount of concessional sales in them also assures that the truly needy in other lands receive some share of the benefits. But if the excess capacity has disappeared from our agricultural sector, then additional exports by us are only going to make a marginal contribution to solving the world food problem. Our perspective has to be substantially broader than just expanding our own output.

Fortunately, we have learned a great deal over the last two decades about the sources of agricultural growth. Unfortunately, we don't seem to be any more disposed now than we were in the past to take the required measures to promote that growth, and in some cases we may even be less disposed. Moreover, the political will to face hard policy choices seems to be as lacking in the developed countries as in the low-income countries.

In speaking to the policy issues, I would like to deal specifically with the problem of population, the problem of production, and the problem of economic policy. And I would like to speak to each of these from an international perspective, ambitious as that may be.

The Problem of Population

Birth rates in the U. S. have declined in recent years to a level consistent with zero population growth, the goal of environmentalists and many other groups as well. That does not mean, however, that our population will automatically stop growing. With the youthful population that we have as a result of high population growth rates in the past, it will be several decades before the population stabilizes. This long delay between the time when the birth rate declines and the population starts to grow at a lower rate is one of the major difficulties the world faces in dealing with the population problem.

The major source of the rapid increase in world population has not been in the advanced countries, however. Rather, it has been in the low-income or underdeveloped countries. And contrary to what many believe, there has *not* been an increase in the *birth* rates in those countries. The increases in population growth rates have been due entirely to reductions in death rates. Moreover, the largest reductions in the death rates have occurred among the young, with the reduction in infant mortality especially notable.

If per capita consumption of food is to increase even modestly in the low-income countries, it seems almost inescapable that they will have to bring their birth rates down. And they will have to bring them down dramatically and quickly. Just to cite one example: If India were to bring its birth rate down to a zero population growth rate this year, there would still be another India by the end of this century. More specifically, India, in another 24 years would have 600 million more people to feed. This almost boggles the mind, especially when one realizes that zero population growth rate for fertility is still a long time in the future.

There are no easy harvests in the population field. To solve the problem requires a great deal of political will, as well as effective organization. We have the technology to make a serious dent in the problem. But the political will is sorely lacking. And unfortunately, solutions to the problem are all too often discussed in terms of population *control*, with all that that implies. In point of fact, in most countries a great deal of progress could be made if we did little more than make available to the low-income groups the knowledge and technology that is available to the upper income groups in those countries. The evidence is strong that these groups, if given that knowledge and technology, will take voluntary measures to reduce their family size. No interference in people's private lives is required.

Similarly, in most countries there are all too many economic incentives that actually stimulate the production of children. Family allowances, preferential tax treatment, and subsidized schooling are just a few examples. The reduction and elimination of these subsidies will also help to lower population growth rates.

The Production Problem

In the short run, major progress will be made in preventing hunger by increasing agricultural output. And it is in this area that we have probably had our greatest advances in knowledge. Research by economists has amply shown that expenditures on research and development are a cheap source of economic growth, as well as a cheap source of increases in agricultural output. Study after study has shown the high social rate of returns to such investments. Yet, in 1974 it is estimated that only $3.8 billion in 1971 dollars was spent around the world for agricultural research. Most of that was spent in the U. S. and other advanced countries. In all the less developed countries (LDC) of Latin America, Africa, and Asia, the total was less than $1 billion (Arndt & Ruttan, 1975).

Considerable progress has been made in this area, however. World expenditures on agricultural research appear to have tripled from 1959 to 1974 —from $1.3 billion to $3.8 billion in constant value terms. And although we lost our will in helping the LDC's on institutional development, there are now some 12 international centers scattered around the world designed to strengthen our agricultural research capability. The improved varieties from a couple of these centers have been an important source of world output expansion in the last decade (Arndt & Ruttan, 1975).

Viewed in its global terms, the major source of output expansion in the future will have to come from the low-income countries. Moreover, that is where the greatest potential is. Yields in those countries are substantially below yields in the advanced countries, and we know that these yields can be increased. The possibility of double and triple cropping is much greater in the low-income countries, since most of them are located in tropical and semitropical parts of the world.

There is one other reason why it is important to increase output in other parts of the world. With a system of floating exchange rates, and with the U. S. open to free trade in agricultural products, the U. S. shares in the problems of world agriculture. For perhaps the first time in our history we therefore have a vested interest in foreign aid. Put somewhat differently, food prices for the U. S. consumer will decline only as world food prices decline. Hence, the U. S. consumer should have an interest in increasing world agricultural output.

This does not mean that we should neglect our own agricultural sector, however. Peterson has shown that the social rate of return to investments in agricultural research is still about 35% in the U. S. (Willis L. Peterson. 1975. Organization and productivity of the federal-state research systems in the United States. Presented at Conference on Resource Allocation and Productivity in International Agricultural Research, Airlie House, Virginia). That is much higher than the average rate of return on ordinary commercial or industrial endeavors, and suggests that in the aggregate society is still under-investing in agricultural research.

In addition to stepping up the rate of investment in our own agricultural research system, there are a couple of other senses in which we could strengthen our science and technology policy. For example, over the years there has been a decline in the relative emphasis on production research. A larger and larger share of the agricultural research budget has been directed to solving ecological, social, and natural resource problems. This was natural in the face of a chronic excess production problem in agriculture, and in the face of the urgency of some of our other problems. But we need to rethink our priorities in the light of the changes in our own as well as the world food situation.

In addition, we have shifted our resource mix in recent years away from basic research towards applied research. The budget pressures many institutions of higher learning faced caused part of this. The desire was to have something that was highly visible and had a quick payoff. These short-sighted objectives have caused us to stress applied research.

This problem was complicated by two other factors. The Federal Government reduced its support for research in general, and an important share of federal money was for basic research. At the same time, there has been a great deal more emphasis on accountability. One of the subtle by-products of this emphasis was a shift towards more applied work, since it is more easily justifiable. My own view is that this emphasis on accountability is going to take the creative life out of our research efforts.

As a final point on research policy, I would like to note that the U. S. academic and research community tends to be terribly parochial. Their emphasis tends to be on their respective state and local communities, with only limited attention given to problems of even a national scope. The international work that we do, needless to say, is of even more limited scope and declining.

The parochial emphasis of our academic and research institutions is somewhat paradoxical. It is in marked contrast to the world view taken by our industrial and business communities, and not at all consistent with the global view of either our foreign policy or our economic interests.

One thing that I fear we may be losing sight of is that over time we make up a smaller and smaller share of the world's stock of knowledge. Although at one time we may have had a dominate hold on the world's stock of modern agricultural technology, it is not at all clear that we do today. Our work abroad should now be done as a basis for learning from others, rather than as a basis for diffusing our knowledge to others. This will require a major change in our thinking and way of doing things.

The Problem of Economic Policy

There is much that I can say on economic policy, given my natural interests. But there are only two points I would like to stress. The first is the complementarity between economic policy, on the one hand, and scientific and technological policy, on the other. No amount of investment in agricultural research will have a payoff to society if the farmer has no incentive to adopt the new technology. This lesson seems difficult to learn. Quite a number of countries around the world are now training a cadre of scientists and researchers. But adjustments in economic policy that would reduce the discrimination against agriculture are far fewer. All too many countries still pursue cheap food policies, where cheap food is obtained by directly lowering domestic prices or by intervening in trade.

The second point I would like to stress is the importance of liberalizing trade in agricultural products. Within the last 5 years, world agriculture has variously been described as being in disarray or in a massive disequilibrium (Johnson, 1973; Hayami & Ruttan, 1971). What the authors have in mind is that an important share of the world's agricultural output is produced in the wrong places by virtue of self-sufficiency and protectionist policies of governments around the world. The corollary, of course, is that if trade were freer we would have a sizable increase in output as a result of countries specializing in what they do best. In the short run, trade liberalization may be the quickest means to increase world agricultural output that we have available to us.

There is another important reason for liberalizing trade. The World Food Conference last November sensitized us to the problem of world food security. Most people construe that to mean the building of reserves and the furnishing of food aid to the LDC's by the advanced countries. Less often is it recognized that liberalizing trade would be an important means of achieving food security. With agricultural products free to move from country to country, a production shortfall in one country could be offset by purchases from a country that was experiencing an increase in production in the same year. Existing barriers to trade inhibit and in some cases prohibit such commodity flows. It is largely for that reason that we may need to return to carrying costly reserves.

A CONCLUDING COMMENT

Even if world population continues to grow at a rapid rate over the next two to three decades, there is no reason why we should have a world hunger problem, or why we should revert to the experience of centuries past when many died from famine. We know how to increase agricultural output, and we know that the potential for the increases in output is there. Just as one example, it has been estimated that the Indo-Gangetic Plain in India has the potential to feed one billion people with existing technology.

The key question is whether we will have the necessary political will. Knowing what to do is one thing. Having the will to do it is quite another. The past records of neither the advanced nor the developing countries would make one particularly sanguine that we will take the measures that are required. Perhaps our best hope rests in a statement by Siegmund Warburg. Warburg said, "Nations and individuals inevitably do the right thing, but only after they have exhausted all other possibilities."

LITERATURE CITED

Arndt, T. M., and V. W. Ruttan. 1975. Resource allocation and productivity in national and international agricultural research. A seminar report, the Agricultural Development Council, Inc., New York.

Economic Research Service. 1974. The world food situation and prospects to 1985. Foreign Agricultural Economic Report, No. 98. USDA, Washington, D. C.

Foreign Agricultural Service. 1975. Foreign agriculture circular (grains). FG-11 75, USDA, Washington, D. C.

Hayami, Yujiro, and V. W. Ruttan. 1971. Agricultural development, an international perspective. The John Hopkins Press, Baltimore.

Johnson, D. Gale. 1973. World agriculture in disarray. Fontana Press, London.

Johnson, D. Gale. 1975. World food problems and prospects. American Enterprise Institute for Public Policy Research, Washington, D. C.

Schuh, G. Edward. 1974. The exchange rate and U. S. agriculture. Am. J. Agric. Econ. 56(1):1-13.

United Nations. 1974. Assessment of the world food situation, present and future. World Food Conference, United Nations E/Conf 64/3.

U. S. Government Printing Office. 1975. Economic Report of the President, Chapter 6. Washington, D. C.

Efficient Fertilizer Use

Stanley A. Barber

Efficient agriculture is important to feed the expanding world population with food produced cheaply enough that people at all economic levels can afford to purchase it. Since fertilizer is a vital component of food production, it is important that fertilizer be used efficiently both to minimize food production costs and conserve natural resources. My discussion of efficient fertilizer use will relate recent concepts on uptake of nutrients from the soil by plant roots to efficient use of nutrients added as fertilizers.

On many of our soils, food production would be greatly reduced if fertilizer were not used. In developing countries crop yields are frequently low because of the lack of fertilizer or inefficient use of the fertilizer that is available. Fertilizer use on a large scale is a relatively recent happening. The large increase in fertilizer use in highly developed agricultural countries has occurred over the last 25 years. Fortunately the fertilizer industry has met the demand by expanding production facilities while keeping fertilizer costs low in relation to the benefit in increased crop yield that results from fertilizer application. As a result, an ample quantity of fertilizer has been the usual recommended method of correcting soil nutrient deficiencies and for supplying the nutrients required to maximize yields. Some producers have fertilized at rates that would ensure that plant nutrients would not limit yields. Increases in all production costs have also stimulated fertilizer use because high yields became necessary to get profitable returns from crop production; high yields are not possible where lack of nutrients limits yields. These high rates of application have not always resulted in efficient use of fertilizer. While use of adequate fertilizer is a relatively recent practice in the highly developed agricultural countries, it has yet to occur in the developing countries.

In the U. S., fertilizer use in 1950, 25 years ago, was only 20% of present use (Hargett, 1975). The average plant nutrient use in the U. S. in 1974 was 102 kg of N plus P plus K for each hectare of harvested crop. This average represents the variation in use from an average 675 kg/harvested ha in Florida to an average 21 kg/harvested ha in South Dakota. Where weather

---

Stanley A. Barber is a professor of agronomy at Purdue University. Contribution from the Department of Agronomy, Purdue Agricultural Experiment Station, West Lafayette, Indiana. Journal paper No. 6049.

conditions have been favorable for crop production, fertilizer use has expanded rapidly over the last 25 years from a situation where lack of nutrients was frequently limiting yields to one where most crops are now adequately fertilized.

The average use of N, P, and K on corn (*Zea mays* L.) in 1974 in the U. S. was estimated to be 115, 30, 76, or a total of 221 kg/ha. The 1972-74 average corn yield was 5,410 kg/ha and this would remove in the grain about 85-15-19 of N-P-K. This is slightly more than half that applied, hence on the soils used primarily for corn production, rates of fertilizer addition to the soil are in excess of the amounts removed in the harvested portion of the crop so that levels of available nutrients in the soil will be increased.

In the countries with a developing agriculture, fertilizer use has increased at a much slower pace so that lack of nutrients frequently limits yields obtained. Nutrient application rates may be less than nutrient removal in the harvested crop so that soils become more impoverished. Recently, the rate of increase in fertilizer use in developing countries has been limited and in some cases reduced because of a rapid rise in the cost of fertilizer. Because capital for fertilizer purchase is often limited in developing countries, efficient use of the fertilizer that is applied is important for their agriculture.

Until recently, fertilizer production and use have expanded at a rapid pace as though the supply of raw materials were inexhaustible and fertilizer would always be inexpensive compared to the value of the increase in crop yield it produced. The last 2 years has brought us to the realization that this may not always be so. We have seen shortages in fertilizer supply and rapid increases in price. Both of these have stimulated people to be more conscious of the necessity of using fertilizers efficiently.

WORLD FERTILIZER RESOURCES

NITROGEN

Nitrogen is present in ample amounts as N_2 in the atmosphere, however, energy is required to convert N_2 to NH_3 in ammonia synthesis plants and natural gas has been used as a source of hydrogen to combine with the nitrogen. Atmospheric nitrogen can also be obtained by symbiotic bacteria that work in cooperation with leguminous species of plants. For future supplies we will need to find a source for the 104 kilocalories of energy required to produce each kilogram of nitrogen in a synthesis plant (Delwicke, 1970) or we will need to develop more leguminous species of plants where the microorganisms fix the nitrogen and supply it to its host plant. Presently much of the energy and hydrogen for ammonia synthesis comes from natural gas supplies. As these supplies are depleted we will need to find alternate sources. The source of energy for nitrogen conversion is the critical factor rather than the supply of nitrogen. Recycling of nitrogen in the crop so that it returns to the soil for use by future crops could be an important factor in reducing our dependence on energy for production of nitrogen fertilizers.

PHOSPHATE

The world phosphate resource that can be economically used has been variously estimated as enough to last from 45 to 450 years depending on the amount of energy that can profitably be used to recover phosphate from the rock source (White & Reynolds, 1974). Much larger quantities of phosphate are present in the world but they require more energy for their recovery for use as fertilizer than is economical at present. Since nutrients are not destroyed by using them but are merely redistributed in a different way, we are in the process of taking phosphate concentrated by nature over millions of years and distributing it over the lands and waters of the world. What will be the phosphorus fertility of soils in the world when we have distributed all that nature has concentrated in a form that we can economically recover? Undoubtedly we would have soils high in phosphorus fertility in many of the present highly developed countries, but would there be phosphorus for economic crop production in those newly developed agricultural areas? It may depend upon how wisely and efficiently we use our resource. While there may be many unknown phosphate resources that will be discovered in the future, we will still need to be concerned with the way in which we use the native phosphate supply because the soil absorbs phosphorus tightly and makes some of it relatively unavailable for future crop use.

POTASSIUM

The potassium resource situation is more favorable (Carpenter, 1975). Deposits in Saskatchewan, Canada alone are enough for the next 2,000 years at current rates of use. Hence the energy required for mining, processing, and distributing the potassium will be the main concern rather than a concern of depleting the resource of the naturally concentrated product.

IMPORTANCE OF EFFICIENT USE

Supply of raw materials, costs of processing, and the need to conserve energy make efficient use of fertilizer more important today than it was in the past. Research programs that have been developed during an era of expanding fertilizer use where fertilizer costs were relatively low have frequently ignored efficiency in fertilizer use. A quote from a recent article on our phosphate supply by White and Reynolds (1974) illustrates the situation.

> The researcher in plant and soil science, bears the responsibility for new discoveries and improved plant technology to increase the efficiency of plants in absorbing phosphorus from that natural sink—the soil. Improving the low plant-use efficiency rate of phosphorus will depend upon increased understanding of phosphate chemistry—from fertilizer plants to crop plants—and progress in this field is imperative as we deplete finite phosphate reserves.

We could also make similar statements relative to the need to use nitrogen and potassium fertilizer efficiently in order to conserve the energy used in nitrogen fertilizer production and to a lesser extent in potassium production. In the last 2 years many articles have been written on fertilizer efficiency. At the American Society of Agronomy annual meetings we had a paper session on fertilizer efficiency and another on genetic control of plant nutrition. It is evident that we are aware of the problem of fertilizer efficiency. Possibly we could gain with a coordinated attack on the problem.

DEFINITION

The term "fertilizer efficiency" used in this paper is defined as the amount of increase in yield of the harvested portion of the crop per unit of fertilizer nutrient applied where high yields are obtained. When changing practices increase the yield response from addition of the same quantities of a fertilizer nutrient, the efficiency of this fertilizer nutrient is increased. It is important to stress that we are particularly concerned with increasing fertilizer efficiency at fertilizer application rates that supply the crop with enough nutrients to get the high crop yields needed to feed the expanding world population. This statement is important because when successive equal increments of fertilizer are added, the greatest crop yield increase usually results from the first increment and the size of the yield increase decreases successively with each additional increment added. Hence the first increment or a low rate of fertilizer application would give the highest efficiency but the yield obtained may be low and unprofitable. This article will discuss research for increasing fertilizer efficiency at high yield levels.

The approach taken in discussing fertilizer efficiency is to look at the basic mechanisms governing the efficiency of fertilizer use by the fertilized crop and discuss ways in which results of present or future research may be helpful in increasing efficiency. Fertilizer efficiency can be considered from two points of view:

1) The efficiency with which a crop plant recovers the fertilizer nutrient that is applied to the soil for use by the crop. This can be use by the immediate crop or a succession of crops. Does the crop recover 5, 20, or 80% of the nutrient applied? Is the remainder lost from the soil or locked up in a relatively unavailable form by the soil so that it is essentially no longer available for use by future crops?
2) The efficiency with which the plant uses the nutrient after it has been absorbed by the root system of the plant. Can we obtain higher yields of the harvested portion of the crop with the same nutrient content?

CURRENT EFFICIENCY OF FERTILIZER UPTAKE

Since we are considering the possibilities for improving the efficiency of fertilizer nutrient uptake it is pertinent to consider the present efficiency of uptake. Only the major nutrients, nitrogen, phosphorus, and potassium will be considered. The nature of crops, soils and practices used will influence efficiency greatly so that the statement made here will be general and refer to situations where fertilizer is used in a reasonably effective manner.

Nitrogen is used extensively for nonlegumes such as corn, cotton (*Gossypium hirsutum* L.), and small grains. Much of the nitrogen absorbed by the plant root is in the nitrate form. That applied as ammonium usually is oxidized to nitrate by the soil nitrifiers. Since nitrogen is mobile in the soil much of it has the opportunity to reach the root surface and the efficiency of uptake is usually high. Estimates of recovery of added nitrogen are of the order of 50% or higher. Hence the possibility of increasing efficiency of nitrogen uptake is a factor of 2.

Phosphorus is immobile in the soil since it is readily absorbed on soil surfaces. It mainly reaches the root by diffusion over short distances (0.02 cm). The amount of phosphorus added as fertilizer that is absorbed by the immediate crop is almost always less than 10%. Succeeding crops will recover lesser percentages. Hence there is considerable opportunity for improving the efficiency of uptake of applied phosphorus fertilizer. Improvement of efficiency of phosphorus uptake is important because phosphate has smaller known world reserves in relation to its need for world crop production than nitrogen or potassium.

Efficiency of potassium uptake is usually between nitrogen and phosphorus. Reliable data are not available because we do not have a suitable radioactive isotope to label potassium fertilizer like we do phosphorus fertilizer. Efficiency of 20 to 40% is probably a good average estimate for uptake by crops such as corn or soybeans (*Glycine max*). The degree of fixation (adsorption of P or K by the soil so that it is no longer readily available for uptake by the plant) of both P and K by the soil greatly influences the eventual efficiency of uptake of these nutrients applied as fertilizers. The type of soil they are applied to and the degree of their reaction with the soil to produce less available forms will greatly influence the percent recovered by the growing crop. There is considerable room for improving the efficiency of use of potassium fertilizers.

INCREASING EFFICIENCY OF FERTILIZER UPTAKE BY CROPS

If we can get the crop to absorb a higher proportion of the nutrients added as fertilizer we will automatically increase fertilizer efficiency. The way to do this will depend on the mechanisms controlling the uptake of a particular nutrient by plant roots growing in the soil. Hence, we will discuss these mechanisms. We have learned a lot about how plant roots get nutrients from the soil in the last 10 years. Most of the nitrogen, phosphorus, and potassium absorbed by roots growing in soil must first move through the soil to the plant root surface before it is positionally available for uptake into the root. The root only contacts 1 or 2% of the soil volume. The proportion of the nutrients in applied fertilizer that reaches the plant root determines to a large extent the efficiency of uptake of these nutrients by the crop. The mechanisms for movement to the root are mass-flow and diffusion (Barber, 1962). Mass-flow is movement of nutrients through the soil to the root in the flow of water to the root that results from transpirational water uptake by the plant. The soil solution contains nutrients which are carried to the root in this flow of water. The amounts reaching the root are determined by the amount of water moving to the root and its nutrient concentration. If this does not supply the requirement of the root, nutrient absorption by the root will reduce the concentration at the root surface and a concentration gradient will result along which ions diffuse to the root. Diffusion results from Brownian motion which moves the ions along a concentration gradient. The nature of the concentration gradient depends on the rate of diffusion of the nutrient in the soil. Phosphorus which is tightly adsorbed by the soil diffuses a shorter distance than nitrogen which is not adsorbed. The general nature of the concentration gradients which may occur radially out from a plant root growing in a silt loam soil as a result of diffusive flow are illustrated for nitrate, phosphorus and potassium in Figure 1.

Mass-flow usually supplies a large part of the nitrogen used because nitrate is present in the water moving to the roots as a result of transpiration. On the other hand little of the available phosphorus and potassium is in solution since most of it is held on the surfaces of soil clay and organic matter. Hence, mass-flow is less important and diffusion is more important for phosphorus and potassium. Because diffusion occurs along a gradient, only those ions that are close to the root will reach the root surface where they are positionally available for absorption into the root. From Figure 1, we see that potassium will diffuse further than phosphate to the root because it has a larger diffusion coefficient in the soil. Hence, a larger proportion of added potassium than phosphorus has the opportunity of reaching the root surface. While values vary greatly with soils and soil moisture conditions, reasonable average distances for diffusion to the root are nitrogen, 1 cm; phosphorus, 0.02 cm; and potassium, 0.2 cm. The mean distance between corn roots in the top 15 cm of soil is about 0.7 cm (Mengel & Barber, 1974), hence some nutrients would need to diffuse half this distance or 0.35 cm before they would become available for absorption by the plant root.

If we assume fertilizer is uniformly distributed through the portion of the soil where roots are active then much of the nitrate can reach the root because it can diffuse further (1 cm) than half the average distance between the corn roots we assumed as 0.35 cm and in addition much of the nitrogen will also move to the root by mass-flow. On the other hand much of the phosphorus is not close enough to the root to be used because it can only diffuse a small part of the 0.35 cm distance. However, phosphorus uptake is facilitated by root hairs that are present on the roots of many plants. Root hair length varies greatly; for corn a value of 0.08 cm is realistic. Since root hairs are longer than the distance phosphorus can diffuse to the root they increase the amount of phosphorus that can reach the root. Assume an average diffusion distance for phosphorus is 0.02 cm adding this to the 0.08 cm root length gives a distance of 0.1 cm about the root that the roots can exploit

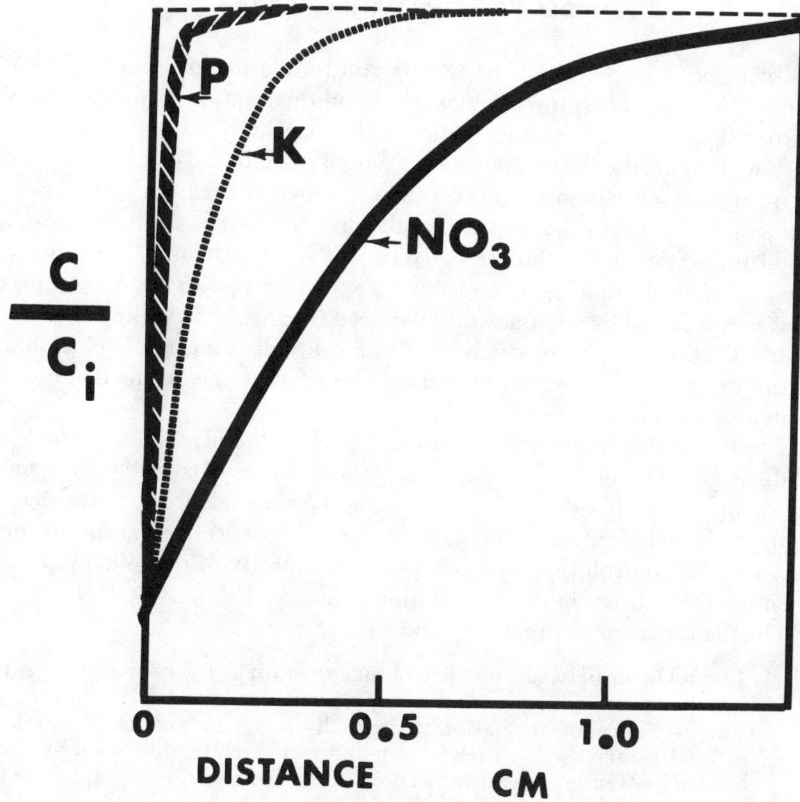

Figure 1. The distribution of NO_3^-, P, and K radially from a plant root growing in silt loam soil, after uptake for 5 days, where diffusion controlled the nutrient flux to the root. C is the resulting concentration as a fraction of C_i, the initial concentration of available nutrients.

for phosphorus. Since this is still much less than half the average distance between roots, phosphorus fertilizer uptake is still not efficient when you consider only the immediate crop. Future crops can, of course, utilize some of this phosphorus provided it does not become so tightly bound by the soil before then that it is virtually unavailable.

Potassium diffuses to the root and the average diffusion distance (0.2 cm) that I chose is also less than half the distance between roots. Its diffusion distance is greater than the usual root hair length so that root hairs do not contribute as much to potassium uptake as they do to phosphorus uptake. The efficiency of uptake is greater than for phosphorus but not so great as for nitrogen. The efficiency of uptake usually reported for nitrogen, phosphorus and potassium agrees rather closely with the distance these nutrients can diffuse to get to the plant root. Hence one reason for low efficiency in uptake is that the distance nutrients diffuse in the soil is small in relation to the distance between roots.

A MODEL FOR EVALUATING UPTAKE

Diffusion is only one of the factors controlling the efficiency of fertilizer use. Nutrients also move by mass-flow and their uptake is influenced by the extent and absorption characteristics of the plant root system. It is necessary to include in the development of a model of nutrient uptake both the plant root and soil parameters. Fortunately progress has been made in developing a model that describes the uptake process in terms of both the root and soil properties that influence the flux of these nutrients into the root.[1] The mathematical model is described in general in this paper and the properties given appear to be the ones that we need to consider. A computer program has been developed to compute nutrient uptake by the model so that only the magnitude of the various parameters needs to be included in order to evaluate their effect.

The parameters influencing nutrient flux into the plant can be divided into those of the soil that influence the flux of nutrients to the root and those of the plant that determine the uptake rate by the plant root system. The principle soil factors are the diffusion coefficient, concentration of the nutrient in the soil solution, and the buffering capacity of the solid phase of the soil for the nutrient in the soil solution phase.

The plant parameters that are important are:

1) The relation between nutrient concentration at the root and uptake rate per unit of root and its variation with root age, plant age, etc. This is a curvilinear relation that usually reaches a maximum uptake rate with increased concentration and can often be described by the Michaelis-Menten equation. An example for phosphorus uptake by corn roots is shown in Figure 2.
2) The rate of water absorption per unit of root.
3) Root radius.
4) The number of roots and their growth rate.
5) The number and length of root hairs.

These parameters describe the morphology of the root and the rate of nutrient uptake by the root and the model allows for their interaction.

Claassen and Barber were able to get reasonable agreement between predicted potassium uptake and observed potassium uptake where as many as possible of the parameters (all for the soil and the first listed of the plant factors) were determined independently of the experiment where potassium uptake was measured.[2] Hence it appeared that the parameters included in the model were the main ones controlling nutrient uptake by the root. This model will be used as a basis for discussing the ways we may be able to improve efficiency of fertilizer nutrient absorption by the growing crop. I will not specifically discuss the common methods of increasing efficiency that most are well acquainted with such as applying fertilizer so that losses by

---

[1]N. Claassen, and S. A. Barber. 1975. A simulation model for nutrient uptake from soil by a growing plant root system. Agron. Abstr. 135.

[2]Claassen and Barber, 1975. Agron. Abstr. 135.

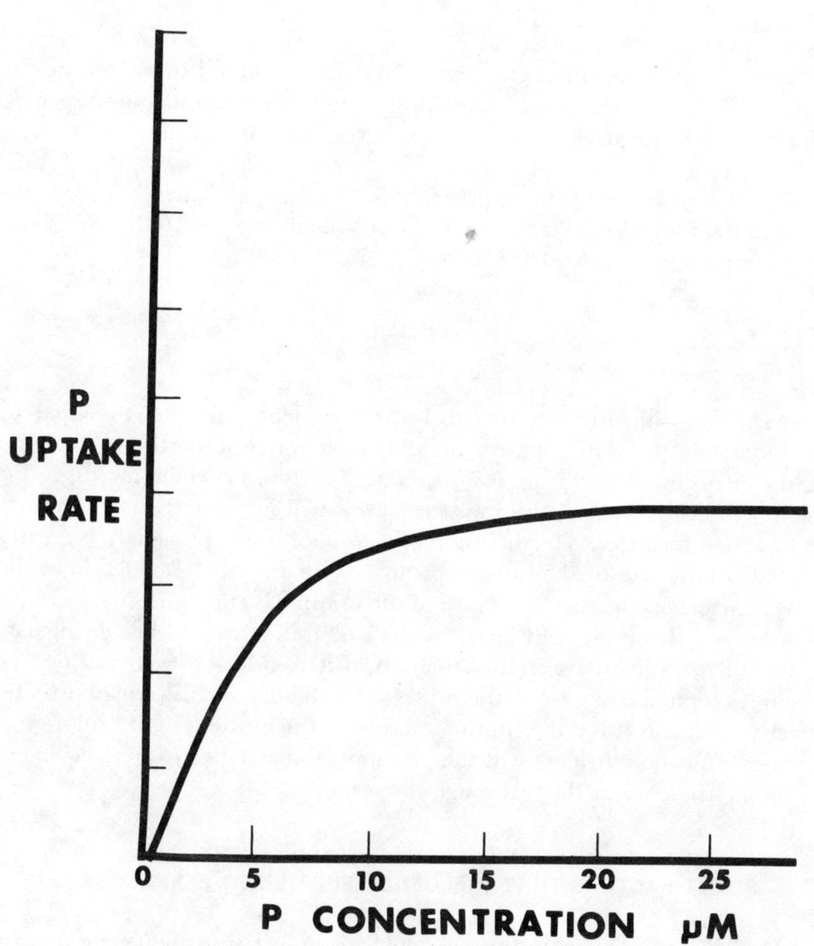

Figure 2. Relation between P uptake rate per unit of corn root and P solution concentration for uptake from stirred solution culture.

leaching or erosion are minimized; that nitrogen is not denitrified or tied up with residue decomposition; that the fertilizer is placed in moist soil that contains actively growing roots; that only the amount needed is applied; that an adequate plant population and optimum row spacing are used; and that irrigation is used to increase yield potential. My discussion will point out areas where research is needed as well as using the results of past research to illustrate how fertilizer use efficiency may be improved.

SOIL FACTORS INFLUENCING FERTILIZER EFFICIENCY

Of the three important soil parameters affecting the rate of supply of nutrients from the soil to the root—diffusion coefficient, nutrient concentration in soil solution, and buffering capacity—diffusion coefficient probably has the greatest effect. In addition the other two parameters also influence the size of the diffusion coefficient. The size of the diffusion coefficient determines how far nutrients can diffuse to the root and for a given spacing of roots in soil it governs the fraction of the nutrients in the soil that can reach the root during the period of plant growth. The size of the diffusion coefficient is affected by three parameters (Nye, 1968). These are the volumetric water percentage, θ; the tortuosity of the diffusion path factor, f; and the buffering capacity, b. The relation is

$$De = Dw\ \theta\ f \frac{1}{b}$$

where Dw is the diffusion coefficient for the nutrient in water. It is apparent that increasing the water content or volumetric water percentage directly increases diffusion. Increasing θ also reduces tortuosity which increases diffusion. Hence increasing soil moisture levels will increase the efficiency of use of added fertilizer. The diffusion of phosphorus and potassium is greatly affected by the size of the buffering capacity of the soil. Reduction of the buffering capacity increases the rate of diffusion. Usually buffering capacity becomes less as the nutrient levels in the soil are increased. Fertilizing part of the soil to a high level rather than all of it to a lower level may be one possibility for increasing De in the soil. One difficulty with altering buffering capacity is that as it is reduced, the fraction of the nutrient in soil solution is increased and the nutrients will then be more subject to loss by leaching in situations where water flows through the soil.

PLANT FACTORS THAT INFLUENCE FERTILIZER EFFICIENCY

Modifying the plant to increase the rate of nutrient uptake per unit of root will increase nutrient uptake and hence increase efficiency of the plant root if the uptake by the plant root rather than the rate of supply from the soil is limiting nutrient uptake rate. Making major changes in the properties of the plant root system is a job for the plant breeder. Some changes in root morphology may be obtained by using tillage and fertilization practices, etc., that change the rooting environment. Can we develop new varieties or develop cultural practices that change the plant root system in such a way as to increase fertilizer use efficiency?

Present knowledge of the nutrient uptake characteristics of plant roots and the amount and size of roots is limited. We do not know much about the variability of these properties that might be useful in studying their inheritance. For corn, we know that uptake rate per unit of root or nutrient influx is greatest with the young plant and decreases with plant age. Can we find cultivars in which this decrease does not occur with plant age? We know that plant roots have a maximum rate at which they will absorb a particular nutrient and that this rate is satisfied at a relatively low concentration of the nutrient in the solution about the root. For example, near maximum uptake of phosphorus by corn roots occurs at a P solution concentration of 0.5 ppm (Jungk & Barber, 1975). Can we find cultivars where uptake rate increases with concentration to higher concentration levels? If we wish to improve efficiency we need the root to be the best possible sink for the nutrient.

Root density in the soil, which is usually given as the length of root in each cubic centimeter of soil, will affect the efficiency of uptake of phosphorus and potassium that diffuse to the root. The more roots, the closer they are together, hence the higher the proportion of the fertilizer that will be close enough to the root to reach it. What controls root size and extent? Is this an inherited property? Can we influence it by changing the physical and chemical nature of the soil? It is apparent that energy spent producing roots will not be available for grain production, however, we could have smaller diameter roots that increase root density without increasing total weight of roots.

Phosphorus uptake appears to be influenced by root hairs and by mycorrhizae. What influences the number and length of root hairs present on roots? What determines the length of time they are active? The rate of phosphorus uptake by plant roots from solution culture is not greatly influenced by the presence of root hairs so the main contribution of root hairs is to supply a greater root surface area for phosphorus to diffuse through the soil to the root. This is important and more research is needed here. If we could get corn roots with lots of long root hairs we could greatly increase the efficiency of phosphorus fertilizer utilization. Preliminary evidence would indicate that root hairs apparently do not affect potassium uptake as much as phosphorus because the average distance potassium diffuses in the soil is farther than the length of the root hairs. Are there differences between cultivars in their ability to produce root hairs? Root hair incidence is affected by the moisture and aeration conditions of the soil. Can we develop cultural practices that will stimulate root hair production and will this increase phosphorus fertilizer recovery?

Some plant species have less root hairs than others; some plants have mycorrhizae in association with the roots that increase nutrient uptake in a manner similar to root hairs. It is now believed that many of our crop plant roots are infected with *Vesicular-Arbuscular* mycorrhizae (Gerdemann, 1974). What influences the degree of this infection? How important is it? Can this be increased so that these fungi will increase phosphorus uptake by the plant roots? The influence of the mycorrhizae appears to be through the increase in surface area of the sink for phosphorus diffusion to the root system.

It has long been believed that root exudates affect availability of soil nutrients. Hydrogen and hydroxyl are exudates of importance and there is evidence that they can increase phosphorus uptake under the proper situation. Hydrogen is released from roots when cation absorption exceeds anion absorption. This can occur when nitrogen is absorbed as ammonium (Riley & Barber, 1971). The hydrogen released by the root reduces soil pH and can increase the P in solution which reduces the buffering capacity, increases the rate of diffusion and hence the amount of P reaching the root by diffusion. When nitrate is the only nitrogen source, anion uptake may exceed cation uptake and the hydroxyl or bicarbonate released will increase the soil pH. In some soils this may increase phosphorus solubility and hence availability, in others it may decrease it.

FLUX PARAMETERS THAT INFLUENCE UPTAKE EFFICIENCY OF EACH NUTRIENT

Since the mechanisms controlling nutrient flux into roots vary with each nutrient, it is useful to look at nitrogen, phosphorus and potassium separately and determine the plant root characteristics that would improve nutrient uptake efficiency.

For nitrogen, mass flow is frequently the major supply mechanism and diffusion is rapid where it occurs. Plant root density in the soil does not have to be high provided roots can absorb nitrogen rapidly per unit of root. To improve efficiency we may need to develop plants with roots that have capability for a high rate of nitrogen absorption and that also can reduce the nitrate level in solution to low levels. A greater abundance of roots would aid in efficiency but it would not be as necessary as with phosphorus or potassium where the rate of diffusion is slower.

Phosphorus moves slowly in the soil so we need plants with an extensive root system and roots that have an abundance of long root hairs. Where root hairs spread throughout all the soil we have the possibility for a high efficiency of phosphorus use. In most instances, with crops like corn and soybeans the rate of flux of the phosphorus to the root rather than the ability of the root to absorb phosphorus limits uptake rate, hence growing cultivars with a large root area for phosphorus to diffuse to and with enough roots so that the average distance between roots is small greatly increases efficiency of phosphorus fertilizer use.

Potassium falls between nitrogen and phosphorus in the rate of diffusion of these nutrients to the root. Since potassium will diffuse further than the common length of root hairs, root hairs will not increase uptake of potassium as much as phosphorus. The root hairs increase the effective radius of the root for absorption of potassium. An extensive system of fine roots that are close together in the soil would be helpful for increasing the efficiency of potassium use.

MODIFYING THE SOIL TO IMPROVE FERTILIZER EFFICIENCY

Since nitrogen is not adsorbed by the soil we can do little to modify the soil to improve nitrogen uptake other than providing a high level of soil moisture for the nitrogen to move rapidly to the root. Ammonium flux to plant roots will be similar to that for potassium since it is bonded exchangeably by the soil in a similar manner.

Modifying the soil to reduce the amount of phosphorus adsorption by the soil, hence reducing the large buffer capacity, will increase the diffusion coefficient and this will increase the amount of phosphorus that may reach the root by diffusion. Modifying soil pH to reduce phosphorus fixation will help by increasing phosphorus level in soil solution faster than the level of adsorbed phosphorus.

Altering the soil to increase potassium flux to the plant roots will also involve decreasing the buffering capacity. Two ways of doing this are increasing the soluble anion content of the soil and increasing the level of exchangeable potassium by adding potassium fertilizer.

COMBINING PLANT AND SOIL EFFECTS

Using our present crop cultivars, we may be able to increase efficiency by the method of fertilizer placement in the soil. Mixing phosphorus or potassium with all the soil frequently results in fixation of a portion of that added and a high buffer capacity. However, restricting the soil contact by banding the fertilizer in the soil, while reducing fixation, will greatly restrict the amount of roots in contact with the added nutrient. From Figure 2, we can see that beyond 15 μM (0.5 ppm) P increasing the nutrient level around a root will have little effect in increasing uptake. A similar situation occurs for potassium uptake. Considering only the plant root, the greatest efficiency in uptake occurs when the added nutrient is distributed uniformly about all the roots rather than concentrated about a small fraction of the root system. This is the opposite to what is needed to minimize adsorption by the soil. The most efficient system may be a compromise. Commonly fertilizer is either applied broadcast and mixed with the soil or banded by the row. These are the two extremes. Possibly an intermediate system would be more beneficial. We conducted experiments with corn where in addition to the two extremes just mentioned we used a treatment where the fertilizer was placed as a strip on the soil surface before plowing so that after plowing it would be mixed with 10 to 20% as much soil as the broadcast treatment. The average yields we obtained over a 5 year period are shown in Table 1 (Barber, 1974). The yield with the intermediate placement of the potassium was significantly higher than for the other two placements. When the corn

Table 1. Effect of potassium placement on corn grain yield and potassium concentration in the ear leaf

Placement	Yield*	K in ear leaf*
	kg/ha	%
Row	7210	1.33
Broadcast and fall-plow	7540	1.37
Strip and fall-plow	8140	1.69

* Values are the average for 5 years results of 3 rates of application, 28, 56, and 112 kg K/ha.

leaf taken at silking was analyzed for potassium its content was also higher. The intermediate degree of fertilizer mixing with the soil had a greater uptake of potassium and therefore greater fertilizer efficiency. The degree that the intermediate placement treatment will increase yields depends on the amount of fixation that occurs in the soil and the relation between fixation and rate of application. When we apply 100 kg/ha of potassium to 20% rather than all the soil, the rate per unit of fertilized soil is 500 kg/ha and usually the percent added that is fixed by the soil is much less. The added fertilizer is contacted by only about 20% of the root system so this may limit the beneficial effect from less fixation and greater rate of diffusion.

EFFICIENCY OF FERTILIZER USE AFTER UPTAKE

Fertilizer efficiency can be increased by getting higher yields with the same amount of nutrient absorbed by the plant. One example of this is the increase in grain yield caused by early planting of corn in the corn belt. When early planted corn is compared with corn planted 4 to 6 weeks later, results such as those shown in Table 2 may occur (Barber, unpublished data). The total dry weight produced per hectare does not change greatly, hence the uptake of nitrogen, phosphorus, and potassium remains about the same. However, the grain yield is higher with the early planting dates so that the corn grain produced per unit of fertilizer used is greater, hence fertilizer efficiency is higher.

Early planting date and higher corn grain yields may also increase the amount of yield response to fertilizer. The data in Table 3 were obtained from an experiment with potassium fertilization of corn planted at four different dates. There was a much larger yield increase with the early planted corn and hence a much more efficient use of the potassium fertilizer.

An example of increased efficiency due to time of application that is most apparent where less than optimum rates are used is the side-dress application of nitrogen. In Table 4 we compare the corn yield increase from 67 kg/ha of nitrogen applied either preplant or as a side-dress application 4 weeks after planting (Barber, unpublished data). The side-dress application is much more effective. With the later application apparently more of the nitrogen is used for producing corn grain rather than corn stalk. This

Table 2. Effect of planting date on yield of corn and proportion of dry matter occurring as grain

Planting date	Dry matter weight: Grain	Dry matter weight: Stalk	Dry matter weight: Total	Grain
	kg/ha	kg/ha	kg/ha	%
April 26	8,060	7,450	15,510	52
May 9	8,460	7,340	15,800	54
May 22	6,940	7,780	14,720	47
June 3	6,160	8,290	14,450	43

Table 3. The effect of planting date on the response of corn to added potassium fertilizer

Planting date	Corn yield		
	Without K	With K	Increase
	kg/ha		
April 26	7,270*	8,900	1,630
May 7	7,280		1,580
May 21	7,070		1,220
June 2	6,550		740

* Values are averages of 9 years data where no K and 93 kg K/ha were applied.

Table 4. The effect of time of nitrogen application on response of corn to nitrogen

Treatment	Corn yield
	kg/ha
No nitrogen	3,010
67 kg/ha applied preplant	5,580
67 kg/ha applied side-dress	7,340

type of efficiency mechanism may be more important in developing countries than in countries with a highly developed agriculture where adequate amounts of nitrogen would usually be used.

CHEMICAL METHODS OF CONTROLLING NUTRIENT FORMS AND INFLUENCING EFFICIENCY

The chemical nature of a nutrient in the soil will influence its likelihood of loss by leaching or by fixation. The most notable example of this is nitrogen which can be present either as the ammonium cation that is held as an exchangeable ion by the soil or the nitrate anion that is almost all in the soil solution and will move with the soil water movement.

The application of nitrapyrin inhibits the *Nitrosomonas* bacteria that changes ammonium to nitrite. When nitrapyrin is applied with ammonium nitrogen, the nitrogen remains as ammonium for a considerable time and is not subject to loss by leaching. Under conditions where nitrogen losses may occur this has greatly increased the efficiency of fall applied nitrogen fertilizers.

REDUCING NITROGEN NEED BY DEVELOPING MORE CROPS FOR SYMBIOTIC NITROGEN FIXATION

Before 1950 legume crops were frequently grown and at least part of the crop returned to the soil to provide nitrogen for a nonlegume crop that followed. This has become an uneconomical practice. The challenge of the future is to make all crops into legume crops so that they will fix most if not all of the nitrogen they require. Recent research (Gamberg et al., 1974) on somatic hybridization in which material is transferred between plant cells has opened up a new area in plant breeding that increases the chances that we will someday be able to develop nitrogen-fixing crops to replace crops such as corn, wheat (*Triticum aestivum* L.), and cotton that now require fertilizer nitrogen.

SOME UNANSWERED QUESTIONS

While we have some of the research information needed to develop crop-soil systems which will utilize applied fertilizers in a highly efficient manner we need new research that will answer a number of questions to enable us to develop additional practices to increase the efficiency of the system. Some of these questions are:

What biological mechanism within the plant regulates the rate at which plants absorb nutrients per unit of root surface?
What regulates the rate and nature of root growth?
What regulates the incidence of root hair growth and root hair length?
What determines the degree of infection of roots with mycorrhizae?
What type of root system do we need on various plant species to give the most efficient uptake?
How should this root system be modified for various soil-climate situations?

If we had the answer to some of these questions and knew the variability of root systems that occur in each species then the plant breeder would be in a better position to develop a cultivar with the type of root system that would optimize efficiency of fertilizer use.

Some questions for the soil fertility researcher are:

How do soil physical and chemical properties influence the extent and morphology of the plant root system?
How can we modify the soil to alter the rates of flux of nutrients through the soil to the plant root?
Can we treat the soil to reduce its buffer capacity for phosphorus and potassium and increase the rate of diffusive flux to the root?

There are many unknowns which should be a challenge to the research scientist. I am sure, with the ever expanding rate of research, that we will answer these questions and develop new technology for the efficient use of fertilizer both in our agriculture and in the agriculture of developing countries where the supplies of fertilizer are more limited and their costs relatively higher in terms of the food produced.

In summary, nutrient uptake efficiency is controlled by the rate and distance of movement of nutrients in the soil, the extent of the plant-root system and the ability of the root to absorb. Practices that increase any of these factors should aid in increasing efficiency of fertilizers for crop production. We need to be aware of the dynamic processes going on in the soil that influence the effectiveness of fertilizer use, and use this information to develop systems for increasing fertilizer efficiency.

LITERATURE CITED

Barber, S. A. 1962. A diffusion and mass-flow concept of soil nutrient availability. Soil Sci. 93:39–49.

Barber, S. A. 1974. A program for increasing the efficiency of fertilizers. Fert. Solutions 18(2):24–25.

Carpenter, J. L. 1975. Potash crisis in Saskatchewan. Fert. Progress 6(2):8.

Delwiche, C. C. 1970. Nitrogen and future food requirements. p. 191–210. *In* D. G. Aldrich, Jr. (Ed.) Research for the world food crises. Publication No. 93. Am. Assoc. for Adv. of Sci., Washington, D. C.

Gamberg, O. L., F. Constabel, L. Fowke, K. N. Kao, K. Ohyama, K. Kartha, and L. Pelcher. 1974. Protoplast and cell culture methods in somatic hybridization in higher plants. Can. J. Genet. Cytol. 16:737–750.

Gerdemann, J. W. 1974. Mycorrhizae. p. 205–219. *In* E. W. Carson (Ed.) The plant root and its environment. University Press of Virginia, Charlottesville, Virginia.

Hargett, N. L. 1975. 1974 fertilizer summary data. National Fertilizer Development Center, Tennessee Valley Authority, Muscle Shoals, Alabama.

Jungk, A., and S. A. Barber. 1975. Plant age and the phosphorus uptake characteristics of trimmed and untrimmed corn root systems. Plant Soil 42:227–239.

Mengel, D. B., and S. A. Barber. 1974. Rate of nutrient uptake per unit of corn root under field conditions. Agron. J. 66:399–402.

Nye, P. N. 1968. The use of exchange isotherms to determine diffusion coefficients in soil. Int. Congr. Soil Sci. Trans. 9th I:117–126.

Riley, D., and S. A. Barber. 1971. Effect of ammonium and nitrate fertilization on phosphorus uptake as related to root-induced pH changes at the root-soil interface. Soil Sci. Soc. Am. Proc. 35:301–306.

White, W., and J. Reynolds. 1974. The phosphate situation: There is no simple solution. Fert. Progress 5(4):8.

Water Management: Some Effects of New Societal Attitudes

Robert M. Hagan

The vital role of water for food production is well known in both rain-fed and irrigated agriculture. As in the past, today's farmers must manage water to obtain both favorable yields and profits. Less well known are the effects on water management of new societal attitudes about the environment, and resource conservation. These new attitudes have already shaped important legislation, influenced the actions of water agencies (for both agricultural and urban areas), and will substantially affect water management at the farm level.

As Dreyfus pointed out in addressing the 1974 American Society of Civil Engineer's (ASCE) National Water Resources Engineering meeting (Dreyfus, 1975), the overriding fact is that water is an essential resource. The clever mind of man has found no substitute for water in supporting all life forms and in producing our food and fiber. Foreseeable demands on water must increase in both diversity and magnitude. Water demands for energy technologies alone will be substantial and necessarily compete with water for agriculture and other uses where water supplies are already over committed, as in the Colorado River Basin. Not only are water resources physically limited in quantity, quality, and occurrence, water use is further constrained by legal and numerous institutional arrangements. The essential character of water resource management in the future will not be the development of new supplies. Instead it will be more intensive *management of relatively fixed water supplies* and, possibly also, reallocation of existing water supplies among competitive uses and users. Supply problems will no longer be mainly a matter of finding reservoir sites and suitable canal alignment. A mix of water uses must be developed that will support the varied goals of society.

Robert M. Hagan is professor of Water Science, Irrigationist in the agricultural experiment station, and extension environmentalist (Land and Water Use), Department of Land, Air and Water Resources, Water Science & Engineering Section, University of California, Davis.

CHANGES IN SOCIETAL GOALS AND VALUE SYSTEMS

Most Americans have ample water delivered into their homes and are so well nourished that they can now afford to be critical of water projects which presently supply their water, food and fiber, industrial needs, some power, and often removal of their wastes.

Some New Concerns

Changing value systems of society now, in many states, result in concerns directed to preserving wild and scenic rivers, shunning dams and interbasin transfers of water as solutions to water needs, emphasizing water conservation, avoiding water pollution, and saving energy. Many citizens today seem much more concerned about environmental protection (with constraints on water use and quality degradation) than in irrigated agriculture—and even prejudiced against any expansion of irrigated agriculture.

In some areas where irrigation is resulting in falling water tables and future water supplies are uncertain some people, even some governmental planners, seem to want the land to go out of production rather than develop new water supplies. Surprisingly, I note this attitude especially among students and younger people in our urban areas. Recently some environmentalists were opposed to developing water for an area now subject to groundwater mining since the more intensive agriculture would increase seasonal farm labor and, hence, welfare costs.

Considerable attention is now being given to water and societal issues. A conference in 1973 at Utah State University on "Societal Well-being—Quality of Life Dimensions in Water Resources and Development" brought together a large group of social scientists to consider social consequences of water development and management.

Several years ago, a group in California circulated a brochure with headlines pointing out that the state would have no water problem if "we would stop giving water to land and give water to people".

Some Responses

These societal attitudes are reflected in the 1973 report of the National Water Commission to the Congress entitled "Water Policies for the Future" (National Water Commission, 1973). The Commission concluded early in its study that its most important task was to reappraise existing water policies and programs in the light of changed conditions and demands and bring them into harmony with the goals of a highly developed, affluent, and urban industrial nation.

The changing scene is reflected by the San Francisco section of the American Society of Civil Engineers in issuing a "Statement of Objectives on

Water Management" which indicates that management must be oriented toward fulfilling all mankind's needs, including improvement of the environment.

In announcing the National Conference on Water held in Washington in April 1975, Secretary of the Interior, Rogers C. B. Morton, chairman of the U. S. Water Resources Council, said, "it is evident that vastly increased demands will be imposed on our water resources to meet national goals. Optimum use and strong environmental protection will require careful planning and management of this essential resource." The Conference's Panel on Water and Food and Fiber called for a national policy on water that "should reflect national goals which specifically address priorities associated with energy production, environmental values, transportation, balance of payments, flood losses, etc. and their relative solutions as compared to production of food and fiber." Concern was expressed that, as more marginal farm lands are called upon to produce food and fiber (to replace production from lands lost to development), numerous environmental spin-offs will occur such as increased erosion, sedimentation, and loss of wildlife habitat. Also discussed were substantially increased water charges to reflect "real" costs and as a means to conserve water. Concern was expressed that regulatory functions arising from the National Environmental Policy Act (NEPA) and the Federal Water Pollution Control Act Amendments of 1972 (PL 92-500) would "have the potential of reducing efficient production of food and fiber and were regarded by some as unrealistic and unattainable." In contrast, the Panel of Water for Municipalities and Industries, in discussing the "zero discharge goals" of PL 92-500, came to the conclusion that "it's good as a goal, but that we do not have all the technology yet to attain zero discharge by 1985 as called for in PL 92-500." The Panel on Water and Food and Fiber closed with a call "to better understand and manage water and to seek answers on environmental quality impacts."

At the Conference, Secretary Rogers Morton called for a national shift in attention from quality of water to quantity of water. He pointed out that the emphasis in recent years has been on concern for our environment and on improving water quality. He continued, "for the next decade and beyond, we will have to expand our concerns to include the cost and quantity of water."

The Panel on Water and Energy pointed out the severe competition expected in western states between energy uses and agriculture and, surprisingly, concluded that water supply is not a problem, but that water management is. The Panel also indicated that institutional arrangements affecting water management are in disarray at all levels of government, leading to indecision. That complaint is heard frequently in California from farmers and agricultural industries seeking to comply with new regulations on water quality protection.

At the first National Conference on "Water Conservation and Sewage Flow Reduction with Water-Saving Devices," held recently at Pennsylvania State University, changes were called for in water consumption patterns, including increased public awareness of the need to "make every drop count." It was pointed out that financial and environmental costs of waste and inefficiency in the utilization of water resources are fast becoming prohibitive. The Keynote Speaker maintained that it is time to abandon the attitude that "water is as plentiful as air" and investigate cost-effective methods of saving water. Although that conference was really addressed to the urban water user, the same pressure will be applied to the agricultural water user.

Water Management Policies

Developments of the past several years are reflected in new water management policies announced in June 1975, by the California State Department of Water Resources. These policies are as follows:

> Water resources of California shall be managed in a manner that will result in the greatest long-term benefit to the people of the State.
>
> Water resources already developed shall be used to the maximum extent before new sources are developed.
>
> All alternative sources of supply, including water exchanges, shall be considered. Conjunctive use of surface and ground water supplies and storage capacity, including planned temporary overdrafting of groundwater, shall be utilized to maximize yield and improve water quality.
>
> To maximize beneficial use, optimum application techniques and processes for water conservation shall be implemented and waste shall be avoided.
>
> Water shall be reused to the maximum extent feasible.
>
> Instream uses for recreation, fish, wildlife, and related purposes shall be balanced with other uses.
>
> Water quality objectives and beneficial uses adopted by the State Water Resources Control Board shall be the basis for water quality management.
>
> Consideration of methods to prevent property damage or loss of life from floods shall include flood plain zoning, flood proofing, flood warnings, and similar measures as well as construction of facilities such as reservoirs and levees.
>
> In comparing alternative water management possibilities, consideration shall be given to capital and annual costs, cost-effectiveness, economic and social benefits, environmental and ecological effects, and energy requirements. The least expensive alternative will not necessarily be selected.
>
> Water management shall be based upon existing laws, but new legislation may be sought where existing law is inadequate.

Effects on Water Management

Developments in the last several years clearly indicate that water management, both in agriculture and cities, is going to be shaped by a host of new societal attitudes, new legislation and regulations, changing water pricing and institutional structures, and other socio-economic considerations. Agriculture will no longer be able to deal with on-farm water management independent of other segments of society. Among other things, agriculture will face a growing competition for water from expanding urban centers and the needs for more power. For example, rice (*Oryza sativa*) farmers in the Sacramento Valley of northern California can no longer manage their water without considering the needs of other beneficial users downstream hundreds of kilometers away—even as far south as San Diego at the Mexican border. To comply with provisions of the Federal Water Pollution Control Act Amendment of 1972 (PL 92-500), they will apparently need to prevent any return flow of irrigation water to streams, although that could impair water rights established to these return flows by downstream users. The latter is one of many complications.

REFLECTIONS IN RESEARCH PRIORITIES

At the recent National Working Conference on Research to meet U. S. and World Food Needs, held in May 1975 under auspices of the U. S. Department of Agriculture and the National Association of State Universities and Land-Grant Colleges, participants sought to identify the problems most in need of research that affect the capacity of the U. S. to meet domestic and international food needs. Study groups examined 49 subject areas covering a wide scope. Of the 49 areas, water was selected as the number 3 priority, with energy number 1 and, interestingly, a specific food crop, soybeans (*Glycine max* L. Merr.), number 2 .

Responding to societal pressures, the current National Science Foundation-Research Applied to National Needs (NSF-RANN) list of Major Areas of Interest (July 1, 1975) gives top priorities to energy and environment. Similarly, a List of Priority Research Subjects for Title II Support in fiscal year 1976 of the Office of Water Research and Technology of the U. S. Department of the Interior includes:

1) improving water resources planning and management,
2) solving energy-related water problems,
3) promotion of water use efficiency, and
4) protection of the environment.

The University of California's Water Resources Center list of research interests for 1975-76 gives emphasis to social, economic, institutional, and policy aspects of water management, water quality and salinity, and water-related aspects of interrelations between agriculture and urban and regional environments.

These evaluations of research priorities indicate the importance of water management and reflect environmental and resource concerns to be considered in developing water management practices.

TODAY'S WATER MANAGEMENT CHALLENGE

The Planet's Limited Water Supply

The steeply ascending curve of the world's population is now thoroughly familiar to agricultural scientists and others. Less well known, however, is the fact that as man's standard of living improves, the per capita demand for water also increases sharply. Thus, if man continues to multiply as projected and to use water as forecast, based on today's trends, the curve predicting future water requirements must ascend even more steeply than that for population. Yet the total supply of water on this planet is essentially fixed. Man must manage his water so as to support "more lives to the gallon".

Water Losses—Challenges for Water Management

The principal processes involved in converting precipitation—through collection, storage, conveyance, and irrigation—into crop production is illustrated schematically in Figure 1 (Hagan et al., 1967b). All elements in this diagram differ so greatly in magnitude that no generalized sketch is possible, yet substantial quantities of water are clearly lost from the watershed enroute to the farm as operational waste, seepage, evaporation, and transpiration. Of the water held by soil within the root zone of crops, generally less than 1% is retained in the harvested crop. More details are given in Figures 2 and 3. The many opportunities for water management suggested by these figures may be summarized as those which: 1) increase inputs of water into the system; 2) collect, store, and transport water from the watershed to use with minimum losses; and 3) convert water at point of use into product with optimum efficiency. Each loss of water into formations from which it cannot be economically recovered, and the numerous losses to the atmosphere by evaporation and transpiration, represent challenges to watershed specialists, planners and operators of irrigation projects, farmers, and soil and water scientists and engineers.

Obviously, these diagrams do not represent all water management problems. Not depicted are water erosion and flood control—important management problems in many areas. For example, as changing public attitudes shift the emphasis from structural to nonstructural solutions to flood control, agriculture will play an important role in flood-plain management, presenting numerous challenges to agronomists.

It is clear that a major challenge to agriculture is the development of water management and other agronomic practices which will achieve more crop per drop of available water. Of all the steps involved in converting precipitation which falls on some distant watershed into crops, many have low efficiency because substantial water losses are involved. The least efficient conversion occurs at the farm in crop production. For some years ahead, water management and other agronomic practices which will increase water

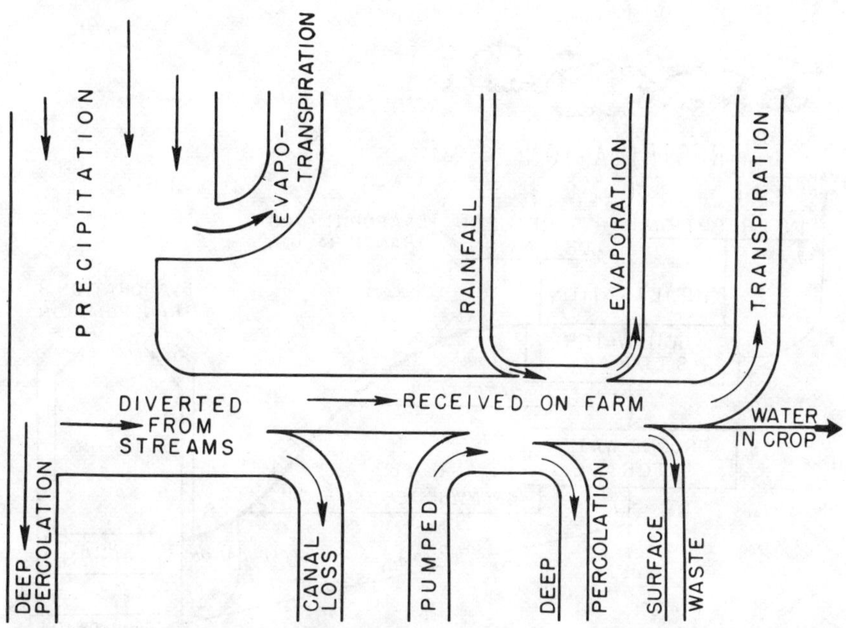

Figure 1. Schematic representation of the principle processes involved in converting precipitation into crop production.

use efficiency at the farm level and deal with new societal pressures, such as land disposal or reuse of wastewaters, will be where the "action is."

In many areas, farmers are becoming aware that, besides managing water to achieve favorable crop yields and profits, they must now be concerned about water conservation. Even in regions of apparently abundant and cheap water, farmers must be concerned with protecting surface and groundwater quality, protecting other beneficial uses of water including fish and wildlife, providing for land disposal and/or use of urban wastewaters, and reducing energy consumption in obtaining and applying irrigation water.

Conditions Affecting On-Farm Water Management

The foregoing gives background for looking at conditions which now greatly affect on-farm water management decisions. Conditions of greatest concern to farmers, at least in the irrigated western states, include:

1) Doubtful adequacy of future water supply arising from pressure to postpone new water projects indefinitely and from increased competition for present supplies.
2) Increased pressure to conserve water through better management and, possibly, cropping pattern regulation.
3) Need for strategy to effectively use less than optimal supplies of water, especially in dry years.
4) Requirement to control and by 1985 eliminate discharge of pollutants to surface water, and probably later also discharges of diffuse flows to groundwaters.

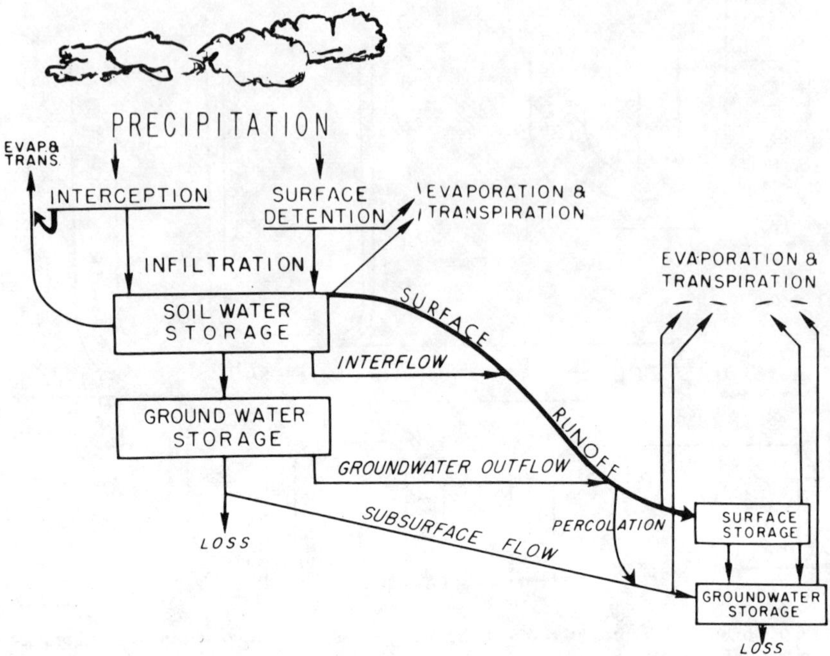

Figure 2. Processes and types of water losses occurring on watersheds, during surface and groundwater flow, and in storage.

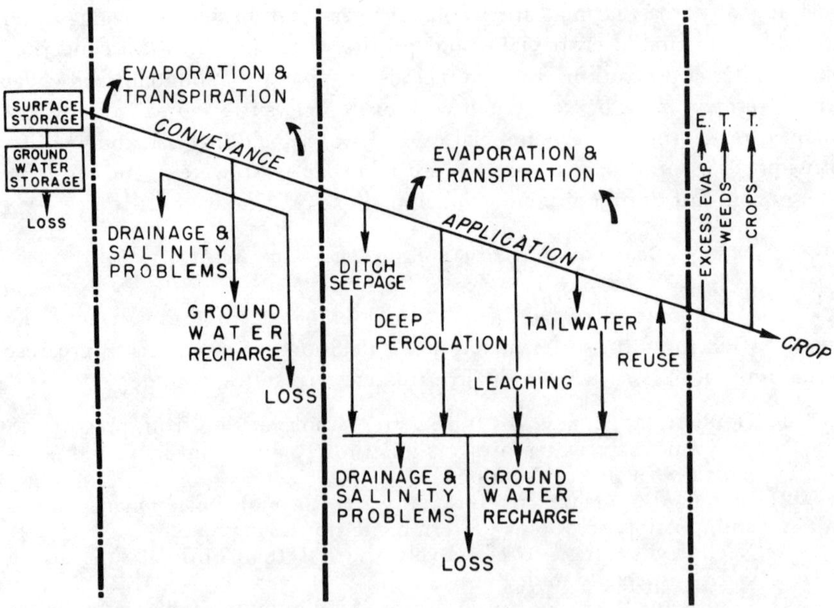

Figure 3. Processes and types of water losses occurring in conveyance, in application of irrigation water, and during crop production.

5) Possible requirement that water must be reused until unusable, to comply with requirements for beneficial use.
6) Increased difficulty in obtaining outlets for drainage waters because of environmental concerns and pollution control measures.
7) Growing pressures to utilize land for disposal or reclamation of wastewaters.
8) Proposals that treated wastewater replace present sources of good-quality irrigation water.
9) Impacts of energy availability, costs, and regulations to use off-peak rather than on-peak power.

Believing these problems to constitute major challenges to agronomists, this paper summarizes the issues and water management problems associated with water supply and conservation, water quality and management, and some energy-water relations affecting agriculture.

WATER CONSERVATION

Today's society substantially agrees that conserving water to eliminate waste is a good thing. This view is clearly shaping programs of water agencies and will soon affect on-farm water management.

Goals and Prospects

This worthy goal, conserving water to eliminate waste, must be approached wisely. Insistence on unsound water conservation measures may, among other things, reduce stream flows, lower water levels in useful groundwater basins, increase salt concentration in both surface and groundwater, allow salt accumulation in soils, reduce per-hectare crop production, increase dollar and energy cost, and, where municipal and industrial waste waters are reused, possibly damage soil productivity or lead to plant pest or human disease problems. In our cities, water conservation through modification of some industrial processes and the toilets and showers in our homes may be beneficial, but some proposals to reduce water use in landscaped areas may be unrealistic and degrade the beauty of the often all-too-limited landscaped areas. In some situations, are water conservation measures necessary or even desirable? It may be pertinent to ask in such cases whether certain water conservation programs will really more fully achieve the numerous and divergent goals of society than present water use patterns? Actual waste of water is undesirable. But I encounter situations, where in our zeal to conform to the conservation ethic, proposals are made which may not in a broad context and in the long run better serve society. This is a difficult area, and generalizations are dangerous. One example may be useful. In some situations, water conservation will save energy, but in others proposed conservation measures will increase energy consumption.

The National Water Commission's Report (1973) recommends specific policies which will conserve water in all of man's activities. It states that substantial savings can be made through improved efficiency in use of irrigation water and suggests greater attention to more efficient irrigation systems, such as trickle or drip. It concedes that almost all irrigated areas need some excess water to leach soils for salinity control.

The Commission correctly recognizes that not all irrigation water applied in excess of consumption use is wasted or lost to the system. It points out that in many cases, perhaps most cases, the surplus water is returned to streams as stream flow or serves to recharge groundwater. Thus the efficiency of water use must be analyzed on the basis of basin-wide or even inter-basin water use rather than in terms of efficiency achieved by each individual user. Even though individual water use may be inefficient, cumulative use in a total system may be highly effective.

In many areas, return flows to streams are important sources of water to downstream users. But, in some cases, return flows will occur later in the year when water is less valuable, or enter the stream below areas of optimum use, or contain added salts which diminish their utility. Water reaching aquifers from which water can be pumped efficiently may provide a useful reserve against future drought. In fact, in many areas overirrigation may be the most effective way of achieving groundwater recharge. But where return flow reaches formations from which the water cannot be economically recovered or reaches a saline lake or the ocean, either as surface or groundwater flow, then improved irrigation efficiency could save water for other uses.

Some other considerations. Water percolating to groundwater basins in excess of minimum leaching requirements may carry excessive salts. Also, dollar and energy costs will be required to pump the water again for irrigation use.

The Commission correctly concludes that improved irrigation efficiency may not "save" large quantities of water in some areas, but may reduce degradation of quality and save energy costs for pumping to recover the water. As the Commission states, "each basin poses its own special conditions and the values gained from better management must be determined by a study of each basin."

Public Image of Water Use by Irrigated Agriculture

Irrigated agriculture has a bad public image. Apparently there is widespread acceptance that farmers are carelessly using and wasting large amounts of the cheap water they have been provided. Seemingly supporting this are figures released by the U. S. Bureau of Reclamation (USBR) indicating that average irrigation efficiencies for surface irrigated fields approximate 44% (U. S. Bureau of Reclamation, 1973. Shut off the water—the root zone is full. Unpublished report). Environmentalists are asking that future water needs be met by withdrawing this "surplus" water from irrigated agriculture. A legislator in California, citing the low irrigation efficiency now achieved and pointing out that sprinklers could achieve 80% efficiency, concluded that presently irrigated land required only one-half the water now supplied.

Dreyfus (1975) at the 1974 ASCE Water Resources Engineering meeting, pointed out that in 1970 irrigation accounted for 83% of the total nationwide consumptive use of water and concluded that "to a very great extent, existing irrigated agriculture makes inefficient use of water supplies." He recognizes notable exceptions but indicates that nearly all projects could significantly reduce losses in conveyance and distribution systems, improve efficiency of water application, and increase productivity of crops.

Water Agency Responses

The U. S. Bureau of Reclamation has been carrying on Total Water Management Studies (1974a) designed to explore ways of getting maximum use of water from present irrigation projects. The Bureau will soon release a new policy basically supporting the concept of more efficient water use. This new policy will apply to new contracts and permits for federally developed water and could, I understand, be made retroactive at the request of individual states.

As mentioned, the California State Department of Water Resources' (DWR) current emphasis is on conservation in use of existing water supplies. Already developed water resources shall be used to the maximum before new sources are developed. This Department has prepared a bulletin on Water Conservation (California DWR, 1976) which describes methods of conservation; estimates possible water savings; discusses relative effects on water supply, water quality, the environment, and the economy; identifies cost ranges; and discusses overall pros and cons. Approaches to improved agricultural water use described include:

1) Efficient utilization of rainfall
2) Seepage control in farm ditches, canals, and reservoirs
3) Cropping decisions including varieties, planting dates, plant population, cropping patterns, and off-season irrigation
4) Reduced evaporation from vegetation, soil, and water surfaces
5) Weed and phreatophyte control
6) Selection of most efficient irrigation method
7) Irrigation scheduling (time and amounts) to optimize crop response and water saving
8) Irrigation systems automation
9) Suitable surface and subsurface drainage
10) Salt management and minimized leaching concept
11) Institutional aspects:
    - Improved water agency organizational and management policies
    - Rehabilitation of antiquated district facilities, including increased water measurement, better structures, etc.
    - More unified planning and administration of water resource development among involved agencies
    - Better integration of water supply, drainage, and flood management systems
    - Reduced water-rights uncertainties
    - Improved water allocation to individual users and more effectively timed water deliveries

Reduced proliferation of special water districts and consolidation of water agencies
Changes in water-pricing policies

12) Artificial groundwater recharge to permit more effective conjunctive use of surface and groundwater supplies
13) Weather modification.

Of the above, the generally more readily attainable and obvious on-farm watersaving practices comprise selecting the best suited irrigation method; achieving good design, operation, and maintenance; and using efficient irrigation scheduling.

New Technologies

Addressed particularly to an international audience, an Ad Hoc Panel of the Advisory Committee on Technology Innovation, Board of Science and Technology for International Development of the National Academy of Sciences (1974) reported on technologies for use and conservation of scarce water supplies. Little-known but promising technologies are described, advantages and limitations indicated, stage of development and research needed, and lists of references and contacts are given.

One new technology in which there has been great interest is drip irrigation, now used on 29,000 hectares (72,000 acres) in the U.S. (California DWR, 1976). The drip method involves frequent low-volume applications of water to soil through emitters spaced along the delivery lines at locations depending upon crop, soil, and other design requirements. Applications must be slow enough to prevent excessive accumulation on the soil surface and yet maintain relatively high levels of soil water within the root zone. Many references on drip irrigation are available from the Second International Drip Irrigation Congress (1974). The water savings possible by this method, where suitable, have led to reports of increasing crop yields with less than half or even one-tenth of the water used under other methods. Such generalizations are for many situations grossly exaggerated and the possibilities of new technologies such as drip should be critically examined before unattainable expectations are built up in the minds of planners and legislators. A few factors to consider in relation to the potentials and limitations of this method include:

1) Possible substantial water savings during establishment period for vine and tree crops where water applications can be limited to the expanding root zone, and the wet soil surface minimized. When the crop canopy approaches 70%, water savings become smaller and limited, mostly to reduced runoff and deep percolation losses.
2) Salt accumulation at periphery of wetted zone and need to control water flow so as to continue water movement beyond main root zone. Also of concern are effects of rainfall in carrying salt into the root zone and ultimate disposition of accumulated salt.
3) Clogging from sediment, chemicals, or biological material.
4) Costs and life of systems.

The California DWR report (1976) estimates that, where conditions are favorable, drip irrigation could save an average of about 35% water over other methods. Enthusiasm for new things is good, but soil and water scientists and other agronomists do have responsibilities to urge realism in projecting water savings achievable by this new development.

Further perspective on drip irrigation is supplied in a recent and thoughtful paper by Rawlins and Raats (1975) which describes the prospects for high-frequency irrigation, including use of drip, solid-set or traveling sprinklers, or small basins periodically refilled with measured quantities of water. Their detailed article concludes that frequent uniform irrigation optimizes the root environment and drastically reduces water use. Considered in their analysis is the view that for most crops, growth proceeds completely unimpaired and crop yield is maximal only when the water potential remains high throughout the life of the crop (Slatyer, 1969; Hsiao, 1973). If keeping plant water potential high gives maximum production per unit area, does it also result in maximum production per unit of water consumed? For considerations of stomatal diffusion resistance, water stress during periods of high potential transpiration, soil water content and hydraulic conductivity, salinity and osmotic potentials in the soil water, water and salt balances of the root zone, the dynamics of the system, and the capabilities of high-frequency irrigation systems, the authors conclude that converting from low to high-frequency irrigation can save water by decreasing deep percolation. Drip irrigation or other methods which minimize the wetted soil area can also save water by reducing evaporation until the crop canopy closes. Information compiled by Rawlins and Raats (1975) suggests that water costs (especially if agriculture had to pay the true cost of water), reduced land preparation costs, decreased drainage requirement, increased yields, and savings in soluble fertilizers could offset much of the cost of high-frequency irrigation systems. Their paper deserves careful study.

Irrigation Scheduling

In contrast with the foregoing discussion much of our irrigated land will probably continue to be irrigated periodically. This brings up the question of when to irrigate. Although a rather time-worn topic, irrigation scheduling still offers an important approach to water conservation on many farms. A tremendous research effort over the years has been directed to this question. Much of this work is summarized in the American Society of Agronomy Monograph, "Irrigation of Agricultural Lands" (Hagan et al., 1967a). It is still impossible to give a simple answer as to when to irrigate, even for a given crop. Efficient scheduling will depend upon many climatic, soil, plant, and management factors, including stage of plant growth and previous water-supply conditions. Electrical resistance blocks, tensiometers, irrigation scheduling guidelines, evaporation devices, and other approaches are available.

Diligent research and extension efforts by M. E. Jensen and others (Jensen, 1969; Jensen & Haise, 1963; Jensen et al., 1970) have led to computerized irrigation reports which provide farmers with information on when to irrigate and how much water to apply for each crop as affected by conditions in each field based on measured or computed evaporation, computation of corresponding evaporatranspiration for the crop, plant growth characteristics including root development, water retention capacity of soil, and other pertinent factors. An Irrigation Management Service based on variations of this approach has operated in Idaho and Arizona since 1969 (C. E. Franzoy, 1969. Predicting irrigations from climatic data and soil parameters. Am. Soc. of Agric. Eng. winter meeting, Chicago. Paper No. 69-752) and in California since 1972 (USBR Irrigation Management Service, 1973. Unpublished Annual Report). Franzoy, from the Arizona experience, estimates that this approach can call for irrigation within ± 1 day when the irrigation interval is about 10 days and within ± 2 or 3 days for intervals of 20 to 30 days. The USBR is actively encouraging water districts to provide or arrange such service for their water users. Guidance in irrigation management is also available from several private agricultural services.

Irrigation with Deficient Water Supplies

Recognizing that less water will likely be available in the future, particularly in the Colorado River Basin and other areas of the west, much attention must be given to operating irrigation agriculture with less than a full water supply. This is a relatively new concept in irrigation management. Present irrigation water requirements and irrigation recommendations have assumed that maximum evapotranspiration will be maintained throughout most of the cropping season, although, in practice, that condition is often not met. Most water-allocation models now being used by planners assume irrigation based on meeting the full evapotranspirational requirements. Forecasts made on this basis conclude that areas of irrigated land will decline in the future. The possibilities of developing water management strategies to optimize crop production from limited water supplies may alter this forecast.

Water Production Functions and Their Uses

To deal with irrigation requirements and to answer other important questions in water-project planning and management requires water-production function research such as now under way by a team of scientists in Arizona, California, Colorado, and Utah in a cooperative project (Office of Water Research and Technology, 1975. Water production functions and predicted irrigation programs for principal crops as required for water resource plannings and increased water use efficiency. Unpublished annual report). Water-production functions relating climatological, soil, plant, and irrigation management conditions provide basic data on which to build reliable crop-production models that can be used to address many vital questions in planning and water management, and hopefully provide information transferable to various locations in the world. The water functions being developed should be useful in addressing topics such as the following:

1) Possibilities for conserving water in irrigated agriculture
2) Simulating effects of different water management programs on crop yields
3) Simulating production from different cropping patterns in accord with different levels of water supply and given soils and climatic conditions. This would include evaluating impacts of deficient water supplies in critical dry years.
4) Planning cropping and irrigation strategies for optimum use of a limited water supply, including selection among crop types and varieties and optimization of hectarages to be planted to different crops
5) Allocating water to agriculture versus other uses
6) Design criteria for water storage, conveyance, distribution, and application systems.
7) Assigning priorities among potential water projects and allocating water supplies among established project areas
8) Planning cropping systems to maximize crop production in areas with small amounts of supplemental irrigation water or full dependence upon rainfall.

Following is a brief summary of water production research by Stewart and colleagues under the OWRT funded regional research project referred to in the beginning of this section.

Water-production function research at the University of California, Davis, is aimed at quantitative prediction of crop yields and optimization of water management in both rainfed and irrigated agriculture.

Yield prediction is based on functional relations between crop production and water at three different levels, as follows:

fundamental relations between dry-matter production and transpiration,
basic field relations between crop yields and evapotranspiration, and
farm-applicable relations between crop yields and water supply.

Optimization of water management is based on determination of:

Differential responses of crops to water deficits in different growth periods, and time-dependent transformations of water supply from all sources (stored soil water, rainfall, irrigation) into evapotranspiration (ET) by the crop, considering crop rooting and leafing characteristics, rainfall and evaporation expectations, soil depth, waterholding capacity and structure, and cropping and irrigation practices.

Examples of the relationships being determined are in the accompanying figures (Fig. 4 and 5), which illustrate findings in field experiments with corn (*Zea mays* L.) at Davis, California in the years 1970, 1971, and 1974.

Figure 4 shows the basic field relationship between percentage reduction in corn grain yield and percentage seasonal ET deficit. A regression line based on the upper portion of the data envelope (blacked symbols) shows the minimum yield loss to be expected from any given seasonal ET deficit. The open symbols underneath the regression line show results when the timing of ET deficits is suboptimal, i.e., when deficits occur too strongly in the more sensitive crop growth periods. Thus prediction of yield is dependent on estimates of both the seasonal ET deficit, and the sequence of deficit times and intensities with respect to growth periods of the crop in question.

The relationships in Figure 4 are transferable between sites where soils and climates differ, provided both are still within the range of adaptation of the crop variety. The reason is that Figure 4 reflects the characteristic responses of the corn variety tested, independent of soil and climate.

Figure 5 shows site-specific relationships between yield increase and applied water, determined at Davis under four different water management treatments, one duplicated with water of higher salinity. The experimental design, developed by Hanks and associates (1974) at Utah State University, consists essentially of a single sprinkler line centered in the plot as the sole source of irrigation. This results in high application rates near the line, diminishing to none approximately 15 m (50 feet or 20 corn rows) to each side. The specific irrigation programs followed were developed jointly by the researchers of the four universities of the Council for International Development (University of Arizona, University of California, Colorado State University, and Utah State University) who are carrying on cooperative research on water-production functions under federal funding.

All treatments included preirrigation to assure that the season began with the soil profile at field capacity. Treatment irrigations began 4 weeks after planting. The remainder of the season was divided into three growth periods, vegetative (to first tassel), pollination (to blister kernel), and grain filling (to maturity). Treatments then consisted of either irrigating to meet all requirements (near the sprinkler line) in any given growth period, or not irrigating at all in that period. Thus in Figure 5, III refers to full irrigation in all three periods, while IOI denotes no irrigation in the pollination period.

In Figure 5 a dashed arrow is projected upward from the lower portion of the curves until it reaches the maximum yield level. This point is labeled ET_{max}, shown by large lysimeter studies to be 673 mm. In effect the arrow represents the regression line seen in Figure 4. Of the 673 mm ET, 421 derived from stored soil water without irrigation, and produced 6,800 kg/ha of grain; the starting point for the five curves. This left 252 mm of ET requirement from irrigation to fulfill the seasonal ET_{max} requirement, and produce maximum yield.

The least irrigation which accomplished this effect was 375 mm seasonal total, administered in 9 weekly applications of normal Davis, California water. This point is labeled IRR_{max}, defined as the minimum irrigation depth associated with ET_{max} and maximum yield, under the specific experimental conditions. Note that seasonal irrigation exceeding 375 mm reduced yield below the maximum, presumably because of reduced soil aeration. These results were produced with the III treatment, which resulted in the highest yields at all seasonal irrigation depths. In this treatment ET deficits away from the sprinkler line were of a chronic nature, occurring all season long at similar levels of intensity.

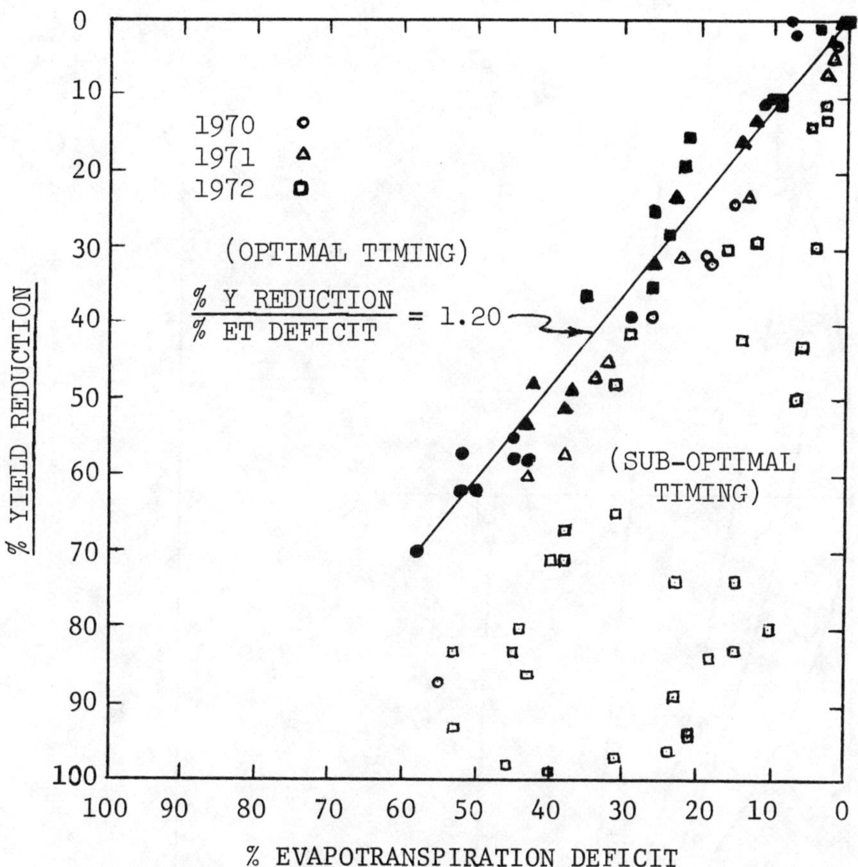

Figure 4. Transferable Y vs ET function for corn showing (a) minimum percentage yield reduction vs percentage seasonal ET deficit and (b) effects of different suboptimal sequences of ET deficits at Davis, California.

This contrasts with treatments OII and IOI, in which acute ET deficits were superimposed during the vegetative and pollination periods, respectively. This finding tends to support the suggestion by Rawlins and Raats (1975) that frequent light irrigations should produce the most crop with the least water when water is adequate, and extends it into the realm of deficit irrigation.

Of additional special interest is the result for the III,S treatment, in which water of the composition projected for the Colorado River in the year 2000 was substituted for Davis water. At the IRR_{max} application rate the gain in yield from irrigation was only 80% of that with Davis water. This corresponds to an increase in ET due to irrigation of 200 mm (see dashed arrow), which is just 80% of the 252 mm of ET contributed by irrigation with normal Davis water. This suggests that increased salinity of irrigation water results in decreased ET by the crop, and correspondingly decreased growth. If so, then the salinity variable can be readily incorporated into all

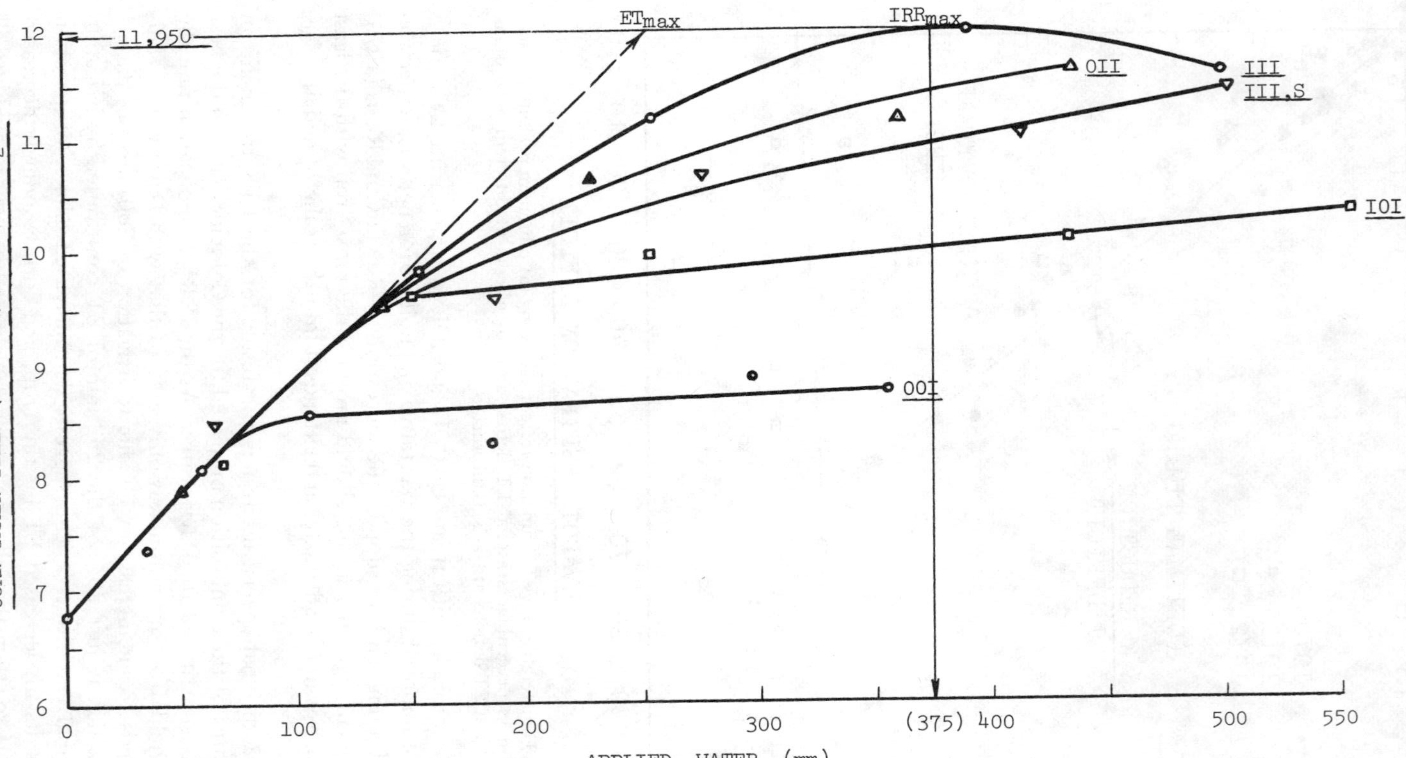

Figure 5. Effects on corn yield of four different irrigation programs which impose varying levels of chronic water deficit in all growth periods, and superimpose acute water deficits in specified growth periods. Additionally a comparison of effects of weekly irrigation with waters of two qualities (III, 850 ppm vs III,S, 1,350 ppm).

relations developed between yield and evapotranspiration. The reader should be cautioned that the findings represented in Figure 5 stem from a preliminary analysis of the data in hand, and are therefore subject to later reinterpretation.

Data on relations between crop yield and water being gathered in the current experiments on the experimental farms of the four Council for International Development Universities will be used to develop one or more models for yield prediction in different water supply circumstances. An example of such a model has recently been published by Hanks (1974).

Water Management for Rainfed Agriculture

The foregoing studies on crop responses to diminishing supplies of irrigation water blend directly into problems of crop production on unirrigated land. Long-term studies, mostly in the Midwest, including effects of depth of soil wetting at time of planting and rainfall patterns on crop yield, have provided useful inputs into water production-function research undertaken by Stewart and colleagues.

Although space does not permit further discussion of this area of research, its importance is clear as farmers seek to maximize their crop yields from available water. If, as a result of constraints on irrigation water supplies in irrigated areas and restrictions on return-flow discharges under water pollution control regulations, the area of irrigated agriculture declines and food production is shifted to rainfed areas as contemplated in the National Water Commission report (1973), water management on these lands will deserve greater attention, especially as additional requirements of water pollution control must be faced.

At the International Crops Research Institute for the Semiarid Tropics (ICRISAT) in India, Krantz and associates (Krantz, 1975) have under way important experiments on water management for rainfed maize, sorghum (*Sorghum bicolor* L. Moench), pearl millet (*Pennisetum glaucum*), and pulses [pigeon (*Cajanus cajan*) and chick peas (*Cicer arietinum*)] . Dr. Krantz believes that water, not land, is the most limiting factor for crop production in the semiarid tropics. His approach is to develop improved land and water management systems which can be implemented during off-seasons. After harvest of the post-monsoon maize, land is plowed, and the seedbed made ready to hold rainwater when it next comes. A ridge-and-furrow system is used and so arranged that any runoff is led down grassed waterways into lined tanks, where it is stored for later application of small amounts of irrigation water at critical times in the following cropping seasons. Using gated pipe or plastic hose, applications of as little as 5 cm of water to post-monsoon crops planted in "relay" (before the monsoon crop is harvested) increased yields of safflower (*Carthamus tinctorius*) from 10.8 to 12.4 and chickpea from 3.9 to 5.8 quintals/ha on red soils (ICRISAT, 1973-74. Annual report). A later report indicates that about a 5-cm irrigation increased yield of maize from 29.6 to 57.5 quintals/ha (personal communication). This striking gain in yield from a small depth of irrigation water suggests a great potential for research seeking optimum water management strategies for use with limited water supplies.

Water Conservation in Urban Areas

In some cities, major efforts are under way to encourage water conservation by industries, public and private institutions, and home owners through educational program, increased water prices, and water use restricttions. In the recently completed report by the California State Department of Water Resources (1976), suggestions are offered to residential users for savings within the home and outside the buildings where an average of 44% of the total residential use occurs in California. Emphasis is directed to reducing watering of landscaped areas, which in California averages 90% of the exterior use. Overwatering of lawns, averaging about 30%, is reported to be waste of water. As with agriculture, not all excess irrigation is lost, for it may percolate to a usable groundwater basin and be pumped and reused. Needless repumping of water wastes energy, however, and may accelerate degradation of the groundwater quality. Where water is imported, this percolating water may be a useful means of groundwater recharge. In other situations, water running off into storm sewers or percolating may be lost to saline bays of the ocean.

Considerable attention is being given to plantings that require less irrigation. Indigenous plants are proposed on the basis they may be less dependent upon irrigation. Drought tolerance or resistance depends on several factors, however, including rooting depth. Planting such materials on shallow soils of some landscaped areas will have disappointing results. There appears to be a need for agronomists and horticulturists to work with urban authorities in developing effective approaches to reducing landscape irrigation requirements while still preserving desired beauty to a reasonable degree. Incidentally, with today's popularity of house plants, the public needs help in avoiding their loss through drought, waterlogging, and salinity.

Water Conservation Through Higher Water Prices

Some urban water districts are actively studying the possibilities of forcing water conservation by substantially raising water prices, and attaching premium prices for excess use and higher rates during the irrigation season. Some planners proposing water-use penalty schedules and prices are completely uninformed about plant water requirements. Without some advice from competent agronomists, some water districts may adopt water-conservation pricing which will lead to neglect of landscaped areas, degrading our urban environment at a time when most citizens want to improve it.

Increasingly, water agencies are being urged to raise prices of irrigation water. The National Water Commission recommends that irrigation water rate structures "be designed to encourage efficient, rather than excessive, water use." It recognizes that water pricing can be valuable to motivating better use, but cautions that it cannot be relied on exclusively to achieve the highest and best use from an overall social standpoint. The structure of user prices should vary from area to area and from situation to situation, depending on conditions. At the recent National Conference on Water, some participants appeared to move in the direction of asking "real" charges for water. Secretary Butz said that putting a realistic price on water might help conserve it but also might result in higher food costs. As the Secretary indicates, a substantial rise in water costs is already on its way as increased project costs, especially for pumping water, are reflected in water charges. As the price of water goes up, there will be greater demands for efficient water management to minimize the cost impact.

WATER QUALITY AND SALINITY MANAGEMENT

Legislation—Requirements and Goals

Environmental concerns led to the Federal Water Pollution Control Act Amendment of 1972 (PL 92-500) passed by nearly unanimous votes in both Houses of Congress. Many states have also adopted legislation seeking to control water pollution. PL 92-500 mandates a sweeping federal-state campaign to prevent, reduce, and eliminate water pollution. The law proclaims the general goals for the U. S.:

1) To achieve wherever possible by July 1, 1983, water that is clean enough for swimming and other recreational uses and clean enough for the protection and propagation of fish, shellfish, and wildlife.
2) To have no discharges of pollutants into the nation's waters by 1985.

As expressed by the U. S. Environmental Protection Agency, these are goals reflecting the "deep national concern about the condition of the nation's waters and a strong commitment to end water pollution." The goals set the stage for a series of specific actions that must be taken—with strict deadlines and strong enforcement provision—by federal, state, and local governments and wastewater dischargers. The law authorizes the federal government to seek an immediate court injunction against any polluters when the resultant water pollution presents "an imminent and substantial endangerment" to public health, or when it endangers someone's livelihood.

The law also requires states to do water quality-control planning in order to qualify for federal grants for construction of municipal and industrial wastewater treatment facilities. California has recently completed detailed Water Quality-Control Plans for 16 hydrologic basins. This planning effort has aroused concerns in irrigated agriculture and illustrated the substantial difficulties to be faced in controlling agricultural return flows in accordance with PL 92-500. It is obvious that many major water management and other problems remain unresolved, particularly the problem of salinity management.

The following requirements on discharges, on an increasing scale, are called for by PL 92-500:

1) By 1977, best practical technology, currently available (BPT CA)
2) By 1983, best available technology, economically achievable (BAT EA)
3) By 1985, "zero" discharge of "pollutants."

In assessing "best practical technology" for a particular category of industry, a balance is to be struck, according to EPA, between total cost and effluent reduction benefits. This could be an important matter in applying PL 92-500 to agricultural return flows. According to EPA, the "best available technology" is the "highest degree of technology proved to be designable for plant-scale operation, so that costs for this treatment may be much higher than for treatment by best practical technology."

Quite precise guidelines for municipal and industrial discharges are set by EPA. The Environmental Protection Agency has not specified BPT or BAT for irrigation return flows, but studies of appropriate guidelines are under way. Permits issued so far under the National Pollution Discharge Elimination System (NPDES) require only the monitoring of certain constituents in return flows, such as sediment, nitrate, and total dissolved solids (TDS).

In brief, PL 92-500 authorizes water quality standards for receiving waters and requires effluent limitations from point sources. In the Colorado River Basin, additional legislation, the 1974 Salinity Control Act (PL 93-320), reinforces water quality control in this basin by authorizing projects to control certain point sources of salinity and provide an important element in the basin-wide plan being developed for water quality and salinity control.

A law enacted in December 1974, PL 93-523 (Safety of Public Waters), supplements the provisions of PL 92-500 by extending controls to groundwater, not clearly covered under PL 92-500. Thus, the degradation of groundwater basins by nonpoint-return flow as percolating water will now come under possible regulation.

PL 92-500 and Irrigated Agriculture

Part of the problem in dealing with PL 92-500 and planning for pollution control is that the law does not clearly define a "pollutant." Various working definitions are in use, and the definition used will affect the acceptability of given water management practices. One definition is that a pollutant is a constituent present in the discharge water at a concentration exceeding its concentration in the supply water. If this definition stands up, it clearly makes agriculture involving transpiring plants a creator of "pollution" by concentrating constituents in water through the loss of pure water from the plant as transpiration.

A source of considerable confusion has been the terms "point" and "nonpoint" sources of discharges. Table 1 summarizes the classifications of point (P) versus nonpoint (NP) as assigned to various types of return flows

Table 1. Point and nonpoint classifications assigned to return flows

Source	Type of discharge	
	Unirrigated agriculture	Irrigated agriculture
Rain Water		
Surface runoff		
Uncontrolled	NP	NP
Controlled	P	P
Percolating water		
Uncontrolled	NP	NP
Controlled	P	P
Irrigation Water		
Surface runoff		P
Percolating water		
Uncontrolled		NP
Controlled		P

arising from rain and irrigation water. For irrigation agriculture, point discharges, which are covered by PL 92-500, include both surface-runoff (tailwater) and percolating water if collected in an under drainage system. Operational spills from conveyance systems are also classed as point discharges.

The Environmental Protection Agency defines point source broadly enough to encompass any discrete surface return flow from irrigated agriculture and, according to a recent federal court ruling, it is interpreted to apply even to small farms. Difficulties in applying the National Pollution Discharge Elimination System (NPDES) to irrigated agriculture include unresolved legal questions, the diversity of irrigation practices, difficulties in distinguishing between point and nonpoint sources of agricultural pollution, the commingling of waters, problems in defining "best practical control technology" for irrigated agriculture, uncertainties about the economic impacts of imposing controls on water management and other farm operations, and numerous institutional complications.

Characteristics of Irrigation Return Flows

Factors determining the quality of return flows from irrigated agriculture may be summarized as:

- Sediments
- Pesticides and other pollutants
- Salt loading
  - Irrigation water
  - Geologic and pedalogic salts in profile
  - Continued weathering of soil minerals
  - Fertilizers
  - Animal and organic wastes
- Salt concentration
  - Consumptive use
  - Irrigation efficiency
  - Leaching fraction.

It is useful to point out the typical differences in the composition of surface runoff and percolating water collected in a drainage system.

SURFACE RUNOFF WATER

May contain higher or lower concentrations of suspended solids than did the applied water.
May pick up fertilizers, pesticides, organic material, and coliform organisms in relatively small quantities that increase with greater erosion.
The concentration of dissolved solids is normally not materially different from that in applied water.

PERCOLATING WATER COLLECTED IN DRAINS

Free suspended solids.
Relatively free of phosphates, pesticides, organic material and coliform.
May contain significant concentration of nitrates.
Will usually contain substantially higher concentration of TDS, with the increase depending on water management practices.

Possible Controls on Irrigation Return Flows

The large number of water management and related practices which could be proposed to implant PL 92-500 include (University of California, Division of Agricultural Sciences, Committee of Consultants, 1974. Unpublished report):

1) Do nothing; no regulation of discharges from irrigated agriculture (quantity or quality).
2) Restrict irrigation in areas where such agriculture causes serious water quality problems (locally or in downstream river systems).
3) Restrict irrigation agriculture to areas where minimum water quality degradation can occur through land use zoning based on suitability of climate, soil characteristics, geology, cropping potentials, and hydrology.
4) Restrict or eliminate irrigation of crops which can be produced elsewhere in unirrigated areas.
5) Regulate amounts and types of fertilizers, amendments, pesticides, and other chemicals used.
6) Dedicate groundwater basins, surface waters, or land areas to receive degraded water.
7) Limit livestock enterprises in irrigated areas to locations which will minimize degradation of surface and groundwater or where discharge can be made to a receiving surface water or groundwater basin set aside to receive degraded waters.
8) Limit development, reclamation, and irrigation of new lands that are salt-affected.
9) Regulate discharges to surface waters based on dilution capacity of receiving water.
10) Export all collectable and unusable discharge water to designated sinks.

11) Provide means of intercepting and collecting water percolating below root zone (as unsaturated flow) and discharge to a suitable sink by installation of an impermeable barrier and placing a drainage system or by raising the groundwater table and installing drainage in the saturated zone.
12) Use drainage wells designed to recover for reuse or export the degraded surface layers of water taken (i.e., skimming wells) to prevent degradation of deeper groundwater.
13) Impose no regulation on irrigation agriculture's discharges but provide for reclamation and dilution, as necessary, of waters later withdrawn in accordance with the needs of specific beneficial uses.
14) Modify, where possible, water quality objectives seasonally as needed to meet seasonal changes in beneficial use requirements.
15) Allow no point source discharges from irrigated agriculture.
16) Require increased efficiency of irrigation to conserve water and reduce total quantity of salt discharged (or mass emission).
17) Require decreased irrigation efficiency to minimize salt concentration in return flows and provide more favorable water for in-stream uses, downstream diverters, and groundwater recharge.
18) Require dilution of discharges to meet the water quality objectives of receiving surface or groundwater.
19) Require the reuse of collectable return flow until no longer usable, only then allowing discharge.
20) Require land, crop, and water management practices which will minimize discharge of sediments to surface waters.

The preceding list of alternatives suggests the dilemma of planners, administrators, legislators, and farmers, who will, in all probability, ultimately have to comply with some combination of the above and still other alternatives in establishing on-farm water management programs.

Management of Pollutants (Other Than Salts) in Irrigation Return Flows

Space does not permit discussion of the control measures already in effect and others proposed to control discharge of animal wastes. Some interesting research has involved analyses of animal wastes as affected by rations. Interestingly, the salt content of waste was found to be influenced by whether the alfalfa hay (*Medicago sativa* L.) was produced on nonsaline or saline soils. The constituents in animal waste were followed as they were carried by percolating waters, a study involving sophisticated research by personnel at the Salinity Laboratory and universities. In areas with high populations of animals, such as dairies and feedlots, animal wastes can contribute substantial amounts of salts as well as nitrate to the groundwater. For these reasons, Regional Water Quality Control Boards in California have proposed to limit the number of animal units per hectare of land over which the wastes can be distributed.

Erosion control has been widely studied and reported. Sediment and other suspended materials in runoff water contribute to pollution of waterways both directly, through the turbidity they introduce, and indirectly, as transport systems for inorganic and organic chemicals adsorbed or otherwise attached to these particles. Since suspended particles constitute the principal carriers for phosphate and pesticides in surface return flows, discharges of these constituents can be largely avoided by preventing erosion. (As indicated above, percolating waters are relatively free of phosphates and pesticides). Space does not permit discussion of the extensive studies of erosion and of water management and other practices that reduce it.

Although nitrogen in soils and waters has been extensively studied, the mass emission of N in irrigation return flows is difficult to estimate. This difficulty is due, in part, to the ubiquitous nature of N in the environment; the diverse sources and sinks; the many pathways by which gaseous, dissolved, and particulate forms of N are transported; and the general lack of an adequate data base for N in return flows.

The ecological impact of nitrogen fertilizers on surface and groundwaters has been the subject of a 6-year study begun in 1972 by the University of California Kearney Foundation, supported by the Research Applied to National Needs Section (RANN) of the National Science Foundation[1]. As summarized in a brief progress report (Coppock & Osterli, 1975), much of the N in soil and water arises from natural sources, more is added as fertilizer, more enters as the result of biological fixation by plants and organisms, and some is added as manure and sewage-plant sludge.

After N fertilizer is applied, it disappears in several directions–into the plant, into the atmosphere, into the irrigation runoff, into soil water within root zone, and into percolation below the root zone. Efficiency in using N varies with many factors, but averages 25-35% for tree crops, 35-50% for vegetables, and 50-75% for forage and grain crops. Thus, typically there will be surplus N available to move with percolating water into drains or to the groundwater. The concentration of N emerging from drains or reaching the groundwater tends to vary with the quantity of N fertilizer applied, efficiency of plant uptake, and extent of denitrification occurring. A detailed sampling program by Pratt and Rible of percolating water under nonagricultural sites and under vegetable fields and citrus orchards illustrates the variable conditions which appear to prevail[2]. They did not find a significant correlation between either the NO_3 concentration or amount of N leached and the amount of N applied. This work and that of others involved in the NSF-RANN study indicates that NO_3 concentrations in the unsaturated soil zone varied from 10 to more than 100 ppm, averaging 56 ppm for 65 sites in southern California. These studies indicate that, at least in some soils, substantial concentrations of NO_3 are percolating toward the groundwater from which, in many basins, water is pumped for both agricultural and urban use.

---

[1]Kearney Foundation of Soil Science, University of California, 1973, 1974, 1975. Nitrate effluents from irrigated lands. Unpublished annual reports to National Science Foundation for Grant No. GI 34733X and GI 43664.

[2]Kearney Foundation of Soil Sciences, 1973, 1974, 1975. Nitrate in effluents. Unpublished reports.

Some potential nitrogen-pollution control measures involving water management follow:

- Farm-level controls
  - Water management practices
    - minimize surface runoff
    - minimize suspended solids in runoff
    - avoid excessive leaching
    - promote denitrification in underdrainage systems
  - Applied N practice
  - Cropping practices
- Project-level controls
  - Water management
    - recapture usable surface runoff
    - increase project efficiency
    - minimize conveyance and distribution losses
  - N removal for nonpoint-return flows
    - promote anaerobic denitrification in drains
    - denitrify in anaerobic deep ponds
    - pass return flow over vegetated wet lands
    - extract N by algae and harvest
    - reverse osmosis, electrodialysis, other desalting processes.

Management of Salinity—Largely as Unresolved Problem

PL 92-500 proclaims as goals by 1983, water clean enough for swimming, other recreational uses, and for protection of aquatic and other wildlife and that will not endanger public health or someone's livelihood. These goals, perhaps, suggest some relaxation with respect to the common salts in water, sodium and calcium chlorides, sulfates, and carbonates, which can reach at least moderate levels before conflicting with the Act's goals. The National Water Commission (1973) similarly recommends that pollution control be directed to protection of designated uses and argues against unneeded reduction in pollution. It calls attention to the diminishing returns as costs for more complete pollution control rise exponentially. These points will presumably be given adequate consideration in setting water quality objectives with which upstream dischargers will need to comply.

Better information is now being obtained on relations between these salts and aquatic life. The U. S. Public Health Service Drinking Water Standards recommend TDS of 500 mg/liter for human consumption. Many estimates have been made on municipal and industrial damage from increased salinity. Municipal damage from salinity in dollars per milligram per liter per year range from the EPA (1973) estimate of \$7,642, USBR (1974b) and Kleinman et al. (1974) estimate of \$119,500, to Valantine's (1974) estimate of \$124,300. The EPA estimate of industrial penalty costs is reported at \$1,148 mg/liter per year, which compares to the USBR estimate of \$1,500.

Damages to agriculture from high levels of salinity arise in three ways: 1) limitations on types of crops that may be grown, 2) reduction in yields, and 3) increased costs to avoid crop losses. Expectedly, damage estimates vary. Damages in dollars per milligrams per liter per year suffered by Lower Colorado River Basin agriculture is estimated by EPA at $45,900, USBR at $108,400 (Po-Chuan Sun, 1972. An economic analysis of the effects of quantity and quality of irrigation water on agricultural production in Imperial Valley, California. University of California, Davis, unpublished Ph.D. Dissertation), and Valantine at $129,300. All estimates include indirect losses to suppliers and processors for agriculture. But Boster and Martin estimate that the damages in Arizona would be less than one-tenth of the above amounts if farmers were given the opportunity to adjust their cropping programs and irrigation management (M. A. Boster, and W. E. Martin, 1975. Economic adjustment to a new irrigation water source: Pinal County, Arizona and the Central Arizona Project. Paper presented to Arizona Academy of Science, Hydrology Section and the Am. Water Resources Assoc., Arizona Section, Tempe Arizona). These few references indicate that the economic costs of increasing salinity are difficult to estimate and depend on many factors. In a recent paper, Robinson et al. compared the effect of water at 877 mg/liter (present salinity in Colorado River) with water raised with the same salts to 1,350 mg/liter (projected future level of river) on yields of 12 crops using sprinkler irrigation (F. E. Robinson, K. S. Mayberry, W. F. Lehman, K. K. Tanji, and J. N. Luthin, 1975. Crop response to higher salinity in sprinkler irrigation. Presented to Am. Soc. of Agric. Eng., Davis, California. Paper No. 75-2061). Five of the crops showed no harm from the higher salinity level. Further studies on damages to various life forms and to municipal, industrial, and agricultural activities would help evaluate reasonable salinity goals.

Sources of Salinity in Water Systems

The salinity of a given water system is the result of both salt-loading and salt-concentrating effects.

Salt-loading effects occur through the addition of salts from natural or man-made sources. This involves the weathering of soil minerals and the dissolving of indigenous residual salt in the soil as rain or irrigation water drains through the soil and overburden materials. This salt-loading process is a continuous process over which man has limited control. Irrigation of previously dry soils in warm arid climates undoubtedly accelerates all processes leading to salt-loading.

Concentrating effects come from diversion of low-salt water out of the basin, by evaporation of water from water and soil surfaces, and by the consumptive use of water by natural vegetation along the waterway and by irrigated crops. The most important of these effects is consumptive use (loss of pure water by transpiration into the atmosphere) with resultant accumulation of residual salt in the root zone. The residual salts are then transported by percolating water to some stream or groundwater.

If the effects of irrigation return flows are to be determined, the contribution of natural lands and their vegetative cover to the salinity of a river system must also be evaluated.

Obviously, there is no absolute measure of the original natural salt load in the Colorado river, but a reconstruction by the USBR suggests that the river in its natural condition delivered 8.4 million metric tons (9.2 million tons) of dissolved solids to the sea annually. The total quantity of salt transported at present has increased only 0.96%. Because of diversions of low-salt water out of the basin and consumptive uses within the basin, however, the salt concentration has increased by 12.6%. More startling are future projections which indicate a further increase of 0.95% in quantity of salt transported, but with a 41% increase in salt concentration.

In general terms, salinity control considerations related to agriculture are concerned with salinity in irrigation return flows. It is assumed that manipulation of the quantity and constituent quality of return flow can be accomplished in two ways: 1) changes in irrigation efficiency, and 2) treatment of return flows to alter salinity. A preliminary report on the Colorado River System by the Utah Water Research Laboratory effectively describes the complexity of problems involved[3]. This report points out that any increase in irrigation efficiency can be considered as a rerouting of water in the system. In other words, the quantities of water following the various flow paths in Figure 6 may be rerouted. The key question is: For a change in ir-

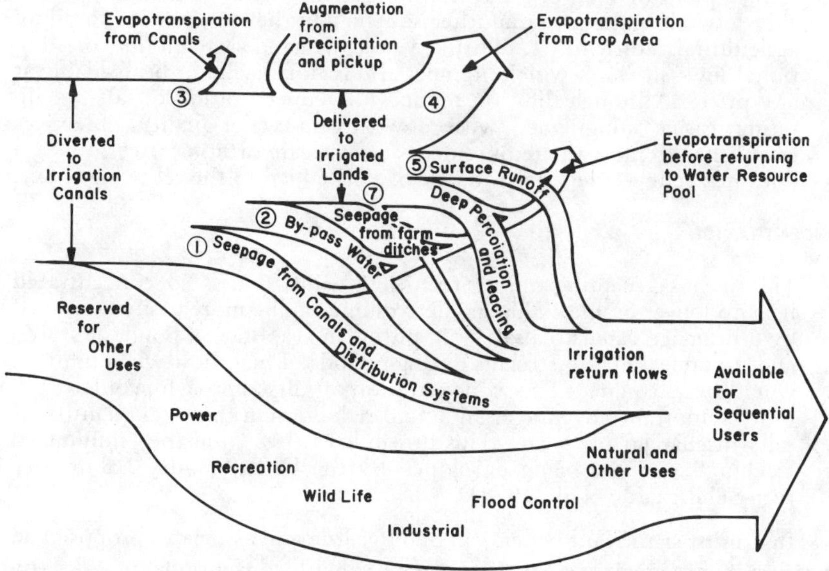

Figure 6. Diagram illustrating the disposition of water diverted for irrigation purposes.

[3]Utah Water Research Laboratory, unpublished report.

rigation practice or pattern of diversion, what are the new routings (quantitative flows over possible paths) and the resulting new salinity concentrations in the stream?

In addition to irrigation-practice options, salinity in return flows could be modified by altering constituents by practices such as: 1) evaporation of water with precipitation and storage of salts in soil profile or water impoundments, 2) dilution of return flow (if lower TDS water is available), and 3) desalting and disposal of salts outside the basin.

Several models have been developed to study alternatives for water quality and salinity management on the Colorado River: a one-dimensional steady-state streamflow and salt-balance model (SALT), a two dimensional hydro-salinity model (BASIM), and a river water-quality model (SSAM) have been developed.[4] Studies employing such models give promise of providing essential evaluations of the many possible water management practices in complex river systems.

The two most obvious water and salt management alternatives are dilution and concentration as described by Ayers and Coppock (1974).

DILUTION

> Dilution of wastewater, a traditional method of water pollution control, may be useful when it makes possible disposal of relatively small amounts of concentrated pollutants such as wastes from a municipal sewage plant or factory. But dilution requires additional water, preferably low-salt water from another source. On the scale needed to dilute agricultural return flows, dilution would demand vast amounts of additional low-salt water which, in most arid regions, will not be available at any price. Although dilution reduces the concentration of salts, facilitating easier compliance with downstream water-quality objectives established to meet the requirements of instream or other uses, dilution does not reduce the total tonnage of salts added to the receiving water.

CONCENTRATION

> This means accepting only return-flow water that is so concentrated it is no longer usable. This smaller volume can be more easily conveyed by a drainage canal to a salt sink, put in an evaporation pond, or stored and later released into streams during periods of high flow when dilution would be adequate. This concentration-of-salts approach may have another important advantage, an actual reduction in the total quantity of salts discharged per year. This possibility arises from the "minimized leaching" concept being developed by the U. S. Salinity Laboratory (van Schilfgaarde et al., 1974).

In a most significant paper, van Schilfgaarde and associates proposed an irrigation management system that, compared with conventional practice, reduces the volume of irrigation water applied, the quantity of salt discharged, and the volume of drainage, while maintaining favorable crop yield. Their approach does not require a salt balance but calls for irrigating with the low-

[4]Utah Water Research Laboratory, unpublished report.

est leaching fraction (LF) commensurate with satisfactory crop growth. Those studies and modeling have verified that minimizing the LF reduces the total salts in the return flow, thereby minimizing surface and groundwater pollution. It maximizes the precipitation of carbonate and gypsum minerals in the soil, minimizes soil mineral weathering and dissolution of salt previously deposited in the soil, and reduces salt pickup from deeper portions of the earth's mantle. Their research shows that reducing the LF from 0.3 to 0.1, can reduce the salt load by amounts ranging from 2.2 metric tons/hectare per year to over 9.0 metric tons/hectare per year (depending on the composition of the irrigation water) without affecting alfalfa yields in the 3-year lysimeter experiment. Table 2 shows the net effect of leaching fraction (LF) on soil mineral weathering less salt precipitation for six rivers expressed as percent of salt input (van Schilfgaarde et al., 1974). These data, expressed as percentage of the salt load of the irrigation water, show that as much as 35% (milliequivalent basis) of the applied salts may precipitate from some waters upon reaction with the soil and that precipitation increases as the LF decreases. However, two of the river waters with low initial salinity (i.e., Feather and Missouri) show a net gain in salt load because of mineral dissolution. This phenomenon is intensified by high LF values. Thus, reducing the LF decreases the salt load of drainage waters because of its effect on both mineral weathering and salt precipitation processes.

As van Schilfgaarde et al. (1974) point out, for this concept to be practical irrigation management systems are needed to provide for closely controlled and metered application rates, uniformly distributed, and with minimal system losses. They recognize, in view of current practices and irrigation efficiencies now generally achieved, both system and farm, that the water management practices suggested are not attainable without major changes. Under the most favorable circumstances, it is doubtful whether an LF as low as 0.15 can be achieved and is more realistically viewed as a theoretical lower limit. Numerous irrigation-system operational problems, water delivery problems, and organizational and legal restraints must be dealt with. Investigators are concerned about resultant changes in the Sodium Adsorption Ratio (SAR) value of the percolating water and also the exchangeable sodium percentage in soil at the base of the root zone. Doubtless there will be problems requiring further investigation. Biggar and Tanji are continuing studies to re-

Table 2. Net effect of leaching fraction on soil mineral weathering less salt precipitation for six rivers

River	Leaching fraction		
	0.1	0.2	0.3
Feather	+158	+243	+344
Missouri	− 11	+ 4	+ 12
Colorado	− 23	− 2	+ 5
Salt	− 10	+ 6	+ 12
Sevier	− 22	− 8	− 3
Pecos	− 35	− 20	− 10

evaluate soil-salt reactions in the light of current interest in assessing and predicting mass emission of salts in irrigation return flows (J. W. Biggar, and K. K. Tanji, 1975. Salt interactions with soil in relation to leaching. Presented to Am. Soc. of Agric. Eng., annual meeting, Davis, California. Paper No. 75-2063). Staff of the Salinity Laboratory and cooperators currently have field trials under way in Arizona and Colorado to explore further the practicality of this promising new approach.

Using present irrigation practices, Robinson (1974) assembled a report evaluating salinity management options for the Colorado River with projected yields for numerous crops in four major agricultural areas.

Disposal of Agricultural Wastewater, Drains, and Sinks

As emphasized in the preceding sections, salt management is generally the major water management problem to be faced by irrigation agriculture in arid and semiarid regions. Salts arising from salt-loading and more important, the concentration of salts by evaporation and, especially, transpiration of pure water out of the system, leave a soil solution higher in total dissolved solids than the supply water. Thus, by common working definition, irrigation agriculture must find a place to store, inactivate, or otherwise dispose of accumulating salts and more-saline waters.

Surprisingly, many environmentalists favor farm water management practices which will hold in some form or precipitate salts in soil and in the earth mantle above the water table. It is pertinent to ask whether this is really the best long-term solution for protection of the ecosystem. The centuries-old experience of irrigated agriculture in the Middle East points out that maintenance of a favorable salt level and achievement of a salt balance is essential to the continued productivity of irrigated agriculture in dry regions.

The foregoing seems to give urgency to finding ways of draining away water that is too salty for crop production and disposing of it in designated salt sinks (including the ocean). Vast amounts of salt are involved. An estimated 18 million metric tons (20 million tons) of salt are now carried by California's waterways. In some rivers they are so diluted, at least during much of the year, as to cause little if any problems. The Sacramento River, a high-quality stream, carries about 110 mg/liter (300 lb/acre-feet) of salts. In contrast, the Colorado River, where water quality is a problem, carries about 740 mg/liter (2,000 lb/acre-feet) of salt at Imperial Dam.

The San Joaquin River, which drains the productive San Joaquin Valley of California, carries a substantial load of salts, much of it from irrigation return flows, and it fails during some of the irrigation season to meet the water-quality objective of 500 ppm TDS as established at the south edge of the Delta. The agricultural wastewater and drain problems of this valley were summarized by Masier (J. W. Masier, 1975. Agricultural waste water in the San Joaquin Valley. Presented at the Am. Soc. of Civ. Eng. Irrigation and Drainage Specialty Conference, Logan, Utah). For years a master drain has been proposed to avoid salination of the San Joaquin River and allow salts to be conveyed through the delta and into the San Francisco Bay system. That has been opposed on the ground that the drainwater, largely because of its content of plant nutrients (mostly nitrogen), could cause algal blooms in the Bay waters, and pesticides, possibly also carried in the drainwater, could harm aquatic life in the Bay waters. A drain serving a part of the Valley and presently terminating in a shallow reservoir has been built by the U. S. Bureau of Reclamation. The USBR, California State DWR, and EPA have jointly studied nitrogen removal by algal growth and harvesting, bacterial denitrification, and a symbiotic process involving algae and bacteria. Removal of N at present-day costs approximates $2.5–8.6/ha-m ($20–70/acre-feet), with some alternatives requiring as much as 12,000 ha (30,000 acres) of land.

A new inter-agency (California State Water Resources Control Board, Department of Water Resources, and U. S. Bureau of Reclamation) study has just begun to explore a range of alternatives, including: 1) the full-length master drain with some treatment of contained pollutants and location of the drain outlet farther into San Francisco Bay to minimize local adverse effects from the effluent; 2) a series of local evaporation ponds to collect and reduce the volume of effluents; 3) use of wastewater to establish brackish wet lands for wildfowl refuges and resting grounds; 4) use of wastewater for power-plant cooling, which will reduce volume but leave a more concentrated blowdown; or 5) some combination of those approaches. Environmentalists, others concerned, and, finally, the decision-makers will have to grapple with the magnitude of the problem—an estimated volume of drainage water for the total valley of about 62,000 to 74,000 ha-m (500,000 to 600,000 acre-feet) per year by the year 2000, and approximately 2.8 million metric tons (3.1 million tons) of salt per year. Perhaps the requirement will be made that return-flow waters from agriculture must be reused until no longer suitable for crop production, with a minimum TDS level set for discharge. Most soil and water scientists appear to believe that the ocean is the logical ultimate salt sink and that transport of salts dissolved in drain water is the most economical in dollars and energy. If wastewaters are used for purposes such as power-plant cooling, some way will presumably have to be found eventually to transport the more concentrated salts to the sea.

Nonpoint Pollution of Groundwater Basins

As required by federal and state legislation, attention is now being directed to nonpoint pollution of groundwater basins by percolating waters. A survey made of the Santa Ana River basin in southern California (Ayers & Branson, 1974), indicated the considerable contribution to the groundwater of animal waste products, particularly nitrates and total salts. Also, samples of water percolating under orchards and other cropped lands collected in the RANN-NSF study (referred to by Ayers and Branson, 1973) indicate substantial amounts of these pollutants are now in the unsaturated mantle below the root zone and percolating toward the watertable.

The Environmental Protection Agency has proposed to set up in the Santa Maria Valley of California a National Demonstration for development of a nonpoint-source groundwater pollution-control program (In-house progress report from the contractor, personal communication). This valley, filled with alluvium containing considerable groundwater now used to supply both urban and irrigation needs, has relatively low rainfall, small river inflow, and a groundwater outlet to the ocean. The goal is to develop water management and other farming practices to protect the quality of this groundwater which can then be applied as a national demonstration. Errors made in planning this project clearly indicate that soil physicists and other agronomists must be involved in these activities. They can have a great impact on water management in the future. This study also raises an important question which deserves careful thought as a guide to future policy and activities toward protecting groundwater quality. Can ultimate salinization be prevented in a groundwater basin which 1) receives limited rain (average 338 mm/year), 2) has small inflow from rivers containing considerable salt picked up from the sedimentary rock watershed (TDS varies with flow and ranges from 1,500 mg/liter at low flow to 200 mg/liter at high flow), 3) has relatively high evapotranspiration rates and dependence on irrigation for production of most crops, and 4) supplies all water used for urban and agricultural uses on the overlying land? It appears that salinization is inevitable unless salt-loading from urban wastewaters is controlled, agriculture consumptive uses are drastically curtailed (presumably by greatly reducing irrigated hectares), low-salt water is imported, or percolating waters are intercepted and desalted an approach involving technical difficulties and, certainly, substantial dollar and energy costs. If it is correct that the practices that have been proposed, such as requiring high irrigation efficiencies, can only slow the degrading of the groundwater quality, attention should perhaps be directed to the possibilities of importing low-salt water and/or allowing gradual salinization of the groundwater, if irrigated agriculture is to continue in this area that produces favorable yields of high-value crops. Situations of this type deserve the active attention of agronomists and other professionals. The long-term implications of alternatives in water management for such valleys should be receiving careful consideration by planners and decision-makers.

Some Critical Problems Remain

The foregoing makes clear that salts arise largely from the natural processes of weathering and that any consumptive use of water, particularly agriculture, concentrates these salts. It is a basic fact of life, not recognized by most people, that plants are water-quality degraders!

If irrigation water is drawn from a given water source, surface or groundwater, and the return flows go back to the same source, "pollution" obviously occurs because of the concentration of salts, and PL 92–500 goals are violated. Thus, growing plants present a serious dilemma in complying with this Act. especially the goal of "zero discharges." This is a vital point which should be understood by the general public and included in the teaching of environmental units in our grade schools.

Agriculturally related pollutants other than salts can apparently be controlled by procedures which may be feasible, at least in many situations. For salts, however, the twin societal goals of water conservation and minimum salt concentration in the nation's waters present another aspect of this dilemma.

Higher water-use efficiency 1) saves water, 2) reduces the volume of drainage water to be discharged, 3) decreases the total salts to be disposed of (reduced mass emission), but 4) increases salt *concentration* in the discharge water–thus increasing the problem of complying with goals to minimize TDS in the nation's waters and with water-quality objectives in receiving waters now set in terms of concentration. Research such as that on minimum leaching now under way by the U. S. Salinity Laboratory is to be encouraged, but not prematurely embraced as the whole solution to the salt problem, as some decision-makers seem inclined to do.

For salt pollution control, the alternative approaches of *dilution* or *concentration* will need to be evaluated in terms of their total impacts, including system interactions, ecosystem reactions, economic consequences, and institutional consequences.

A detailed report,[5] not yet released, lists a number of questions and impact issues resulting from PL 92-500 in connection with salinity control. With respect to salt loading, it raises many questions including:

Will discharge permits induce less efficient irrigation to raise the leaching fraction?

To what extent will discharge permits accomplish effluent limitations, or will they be counter-productive?

Will water-quality limitations on agricultural return flows lead to changes in the concept of beneficial consumptive use to require more advanced technology and management practices?

To what extent will practices ultimately called for under best practical technology (1977) and best available technology (1983) modify salt loading?

What will be the effects, including environmental, of using the soil as a salt sink?

Another very pertinent question is raised by the National Water Commission (1973) in asking whether harm to air and land from excessive water pollution control may be "far more damaging to the environment than the retention of some small amount of pollution. . .and where other uses of water are not adversely affected."

These, and many additional questions raised, will surely call for careful investigation by professionals in many fields, public leaders, and decision-makers. The future pattern of farm water management, especially in the western states, will be decidedly affected.

[5]Utah Water Research Laboratory, unpublished report.

The societal attitudes leading to proposals that farmers in western states increase their irrigation efficiency and avoid return-flow discharges containing objectional concentrations of pollutants create problems for today's society that are practical and difficult to resolve.

Increased irrigation efficiency to reduce water pollution not only reduces return flows to possible downstream users, but the return flows will be more saline.

Although it is often said, "dilution is not the solution to pollution," dilution (in the minds of many) is still an obvious approach to reducing the pollutants, especially the salts, in our waters. However, in situations like the Colorado River Basin, rivers are already overappropriated. In other words, there is no water to which someone does not already have a right, and hence no water is available for dilution, as there will often be in the more humid areas.

Along some important rivers, many existing rights to divert irrigation water depend upon return flows from upstream diverters. Pollution control achieved by recycling, evaporation, or treatment and reuse by upstream users thus injures junior water rights. If increased consumptive use arises from pollution-control measures, such use may be enjoined by a junior appropriation thereby deprived of water. The only alternative available to the senior who has increased his consumptive use (inflow minus all outflows) is to acquire sufficient water with senior rights to make up the additional consumptive use, or to reduce his initial use to the extent that his new total use plus his increased consumptive use will equal his historic consumptive use. All of this may cause serious dislocations of existing water uses, or, in the alternative, substantially reduce the initial beneficial nonevaporative uses. The foregoing illustrates, in a relatively simple case, some of the difficult legal questions to be faced by the water manager seeking to control pollution.

From an economic perspective, several types of policies are proposed to induce improvements in water quality, including: 1) providing additional quantities of water for dilution, 2) public investing to improve quality, 3) initiating litigation against major polluters, 4) imposing strict water-quality standards for all users, and 5) implementing direct economic incentives—probably either taxes on pollution or subsidies for water-quality improvement.[6]

Social, Economic, and Environmental Impacts of PL 92-500

Concerns of Congressmen about the impact on the country of such sweeping legislation as PL 92-500 led to provisions to establish a National Commission on Water Quality. It is charged with making a full study of all technological aspects of achieving the effluent limitations and goals of this Act—and also all aspects of the total economic, social, and environmental effects of achieving or not achieving them. The Commission is to report the

[6]Utah Water Research Laboratory, unpublished report.

results of its investigations, together with recommendations to the Congress. The report is due early in 1976.

Numerous types of studies are under way. These include national technology studies of the basic industries (including irrigated agriculture), social and economic studies, environmental studies, institutional studies, attitudinal studies, and regional assessment studies of impacts in selected river basins and waterway systems. Several studies are exploring various possible scenarios and their various impacts on society.

The technology studies will inquire into the degree of point-source effluent reduction, in terms of both volume and constituents, achievable through implementation of this Act. Methods of minimizing pollutants from nonpoint sources will also be analyzed. In the environmental studies, the Commission will identify the chemical, physical, and biological composition of water necessary to provide for aquatic life and recreation. An analysis will be made of the capabilities of technology to reduce or eliminate discharges of pollutants, and the resulting water quality expected will be compared with the level necessary to support the 1983 requirements. Because there will be residuals from some effluent reductions, the environmental effects of their disposal will also be considered.

Major attention is being given to analyses of all aspects of projected economic impacts, both positive and negative. Grave concerns about the costs of pollution abatement had been raised by the National Water Commission (1973) which particularly criticized the "zero" discharge requirement as impractical and unattainable. Similar concerns are shared by agriculturists. As one example, the USBR estimated that the cost of processing drainage water in the San Joaquin Valley of California to remove nutrients and salts to comply with expected discharge requirements would probably exceed $18.60/hectare-water ($150/acre-feet) in an area where water costs are now $0.93/hectare-water ($7.50/acre-feet). Hopefully, the Commission's study will provide useful guidance to the Congress in this important matter which will so substantially affect water management and vitally affect irrigated agriculture and many other water users.

WASTEWATER TREATMENT, DISPOSAL AND/OR REUSE

Wastewater Treatment

With the rising costs and also the considerable energy demands of secondary and tertiary treatment plants needed to meet the discharge requirements of PL 92-500 and 93-523 (Safety of Public Water Systems Act), there has been considerable interest in disposing of wastewaters on land areas.

This concept has been with us for centuries, but the new requirements of today and the future pose new questions about its suitability for proposed widespread and varied use. Research by EPA on land treatment of wastewaters has recently been summarized (EPA, 1975). Three approaches are considered 1) infiltration-percolation, 2) overland flow, and 3) irrigation (discussed in a later section).

Infiltration-percolation studies at several locations in Arizona, California, Minnesota, Wisconsin, and New York are reviewed by EPA. The design and operation of these systems have emphasized removal of nitrogen and other pollutants in treatment-plant effluent. The effectiveness of removing biochemical oxygen demand (BOD), nitrogen, phosphorus, and pathogens depends on factors such as soil, infiltration rates, and relative periods of spreading and intermittent drying. Although the two north-central locations represent radically different climate, overall performances were similar to that in the southwest. EPA concludes that technological data are now available to design and operate systems for a limited number of situations, but of more importance is the apparent utility of the approach under widely differing climates.

The Corps of Engineers has conducted additional studies on the percolation method as an extension of the Pennsylvania State University "living filter" concept, using several design alternatives for handling input wastewater and land-treated effluent (U. S. Army Corps of Engineers, 1972). These included direct groundwater recharge by the percolating water, percolation followed by lateral movement to a surface water, or pumped withdrawal, usually from a system of underdrains.

Overland-flow treatment of municipal wastewaters is a newly developing technology. This method is being used successfully for treatment of cannery wastewater in California. An EPA report indicates that this overland method may even be satisfactory for raw sewage, and researchers hope to develop overland-flow for treatment on a year-round basis of raw sewage from rural communities to standards equaling secondary treatment but at a cost approaching primary.[7] Here there will be need for agronomists' knowledge in establishing plant covers under these excessively wet conditions.

A recently issued EPA Technical Bulletin provides an evaluation of land application systems with considerable details on wastewater management plans, design plans and specifications, and operation information (EPA, 1975). Useful appendices include 185 references and a selected annotated bibliography.

Once-Through Versus Multiple Use of Wastewater

PL 92-500 encourages wastewater management with revenue-producing facilities. Although some sections of this Act are aimed at stimulating the development of reuse projects, the law has a number of constraints which inhibit such projects. California's water quality-control act states that all possible steps be taken to encourage the development of facilities for both water reclamation and its reuse. In accordance with this, one of the key policies of

[7]R. E. Thomas, and C. C. Harlin, Jr., 1975. EPA research on land treatment. Prepared for U. S. EPA Technology Transfer Program Seminar, San Francisco, May 1975.

the California State Department of Water Resources is the reuse of wastewater "to the maximum extent feasible" before consideration is given to developing new sources of water.

Water supplies may have a once-through use or be circulated for multiple use within systems of various sizes. The reuse system may consist of one farm, one city, a whole river basin, or even several interconnected basins including groundwater basins.

Much of the nation's water has been reused for years, either in passing from one city to the next, from one field to another within a farm, from one farm to the next, from farm to city, or more recently, from city to farm. It is the last possibility that is now receiving much attention.

Once-through use is being increasingly criticized. For examples, cities of the San Francisco Bay region import high-quality snow-melt water from the Sierra Nevada mountains by aqueduct. This water is typically used once by homes and industry, collected in sewers, treated as required, and discharged into saline San Francisco Bay. The East Bay Municipal Utility District has been sued by the Environmental Defense Fund to force reuse of this water, presumably for industry within its service area, on the grounds that present once-through use does not adequately meet the requirements for beneficial use. Feasibility studies have been made of collecting wastewater from this metropolitan area and exporting it for use in augmenting stream flow out of the Sacramento-San Joaquin Rivers Delta for salt repulsion or irrigating croplands in the Central Valley.

Stemming from pressures to maximize use of already developed water supplies, one water district in California has been informed that it must plan for reuse of its wastewater before a USBR project to import water will be supported.

What alternatives exist for reusing urban wastewater? Generally these include 1) groundwater recharge, 2) industrial processes (cycling within plants or return of treated wastewater to industrial plants), 3) power-plant cooling, 4) irrigation of parks, golf courses, roadside plantings, and other large landscaped areas, or 5) irrigation of farm land.

Groundwater Recharge

Groundwater recharge with wastewater is not a new concept, but both federal and state legislation require protection of groundwaters from pollution. Where recharge involves aquifers which serve as the supply of domestic water, recharge criteria have not yet been clearly established. Still to be resolved are water quality requirements, public-health constraints, institutional arrangements, and, in some areas, social concerns. There is still great resistance to use of reclaimed water for household purposes. Thus, there is reluctance to see wastewater used to recharge aquifers from which their water comes.

Industrial Use and Power Plant Cooling

Industrial use will have varied requirements, and power plant cooling may require reduction in such constituents as N and P, and possibly reduction in total dissolved solids. A number of agencies, including the USBR, California DWR, power companies, and universities, are seeking to develop technologies to increase the usefulness of agricultural return flow by reducing contained plant nutrients and salts. Two major power plants are now planning to use agricultural return flow in California's San Joaquin Valley to the extent that such water will be available.

Reuse of Urban Wastewater for Irrigation

This is one of the most active areas of water management. Numerous papers at the American Society of Agronomy annual meetings dealt with aspects of this problem.

Environmental and resource conservation concerns, described earlier in this paper, have combined to popularize the idea of reusing urban wastewater for irrigation. For example, in Sonoma County, California, rapidly growing cities expanding onto the limited surrounding agricultural land and also facing a requirement to improve their wastewater treatment before discharge into a waterway, were urged by a coalition of environmental groups to consider, instead of building a tertiary wastewater treatment facility as now increasingly expected, using wastewater after secondary treatment to irrigate land around the city, thus preserving a green belt against further urban expansion, preserving open space, and producing feed and forage crops locally to benefit the financially hard-pressed dairies which, if they were to close, would leave more now-open land for urban development. The environmental groups got the sanitary engineers involved in planning new sewage treatment facilities and the officials in the State Water Resources Control Board and its Regional Board, which sets discharge requirements, to agree to expanding agronomic trials, already under way by the Cooperative Extension Service, to determine the feasibility of such an approach.

Under the provision of PL 92-500 and policies concerning making federal and state grants available for sewage treatment facilities, cities and water district pressures to use urban wastewaters for irrigation within the cities or on surrounding land either cropped, planted to forest trees, or used for feed or green fire breaks on watershed lands.

While the idea is attractive to many people today, there are many real problems, including:

1) Location:
   Many major sources of urban wastewater are distant from land areas on which water could be effectively used.
2) Relatively small quantities of wastewater:
   Even large metropolitan areas produce volumes of wastewater relatively small in relation to the large quantities of water required for irrigation in arid and semiarid regions.

3) Seasonality in need for irrigation water:

In many areas the irrigation season extends for only a few months, whereas wastewater is produced in nearly equal amounts throughout the year.

4) Need for seasonal storage of wastewater or alternative means of disposal during nonirrigation season:

This may require a storage reservoir or a more refined wastewater treatment plant, such as tertiary, in order to comply with discharge requirements when not used for irrigation. In some cases, higher wintertime river flows may allow such discharges.

5) Relatively high cost:

Costs are high for the above reasons, plus conveyance and distribution canals or pipelines.

6) Water quality:

Will contain N and P, which may usually be desirable except on some crops near maturation.

Wastewater may contain heavy metals, which can often be reduced within the city by source control, but whose effect on soils and crops needs further research. Limited experiences from heavy metals concentrated in vegetable crops point to a need for caution and research.

Will contain higher TDS than water intake to city by about 300 ppm. Resultant wastewater will thus contain more salt than present irrigation water if drawn from the same source, but TDS may be lower where irrigation water is drawn from a source with higher initial TDS.

Several guidelines for interpreting water quality for agriculture have been prepared (including NAS-NAE, 1972; R. S. Ayers, and R. L. Branson, 1975. Guidelines for interpretation of water quality for agriculture. Prepared for University of California Committee of Consultants, University of California Cooperative Ext. Serv. Mimeo.; J. E. Christiansen, E. C. Olsen, and L. S. Willardson, 1975. Irrigation water quality evaluation. Am. Soc. of Civ. Eng., Irrigation and Drainage Specialty Conference, Logan, Utah). Predictions of long-term effects of using given wastewaters on soil productivity and composition of crops grown are different because many factors are involved, including climate, soil, crop, irrigation system and its management, and drainage conditions.

7) Public health:

Public health departments restrict the use of treated municipal wastewater on edible crops. Considerable uncertainty exists as to any real hazards, and this matter is being actively explored. Viruses and organics are of concern. The Occupational Health and Safety Act may also impose restrictions on workers in areas irrigated with wastewaters.

Those pressing for wide-scale reuse of wastewater for irrigation will need to recognize that, although the concept is attractive, many problems remain. The wastewater, even though it contains nutrients (mainly N and P) which the farmer will usually be purchasing in fertilizer, will not represent an unmixed blessing to the user. Hence, urban dwellers, who have a waste product which must be disposed of, will likely have to share some of the costs of making this water attractive to irrigators if they have other sources of water of better quality, free of restrictions, and of lower net cost.

Reuse of urban wastewaters for irrigation will likely be most successful in the southwest where the rainfall is low, the growing season long, the soils low in nitrogen, the basic or alkaline soils can inactivate any toxic metals in the effluent, and water supplies for irrigation are scarce and increasingly expensive (G. V. Johnson, 1974. Utilization of sewage effluent for irrigation of turfgrasses in arid regions. Environmental Considerations in Turfgrass Management. Symposium at the 1974 ASA meetings in Chicago, Illinois). However, Thomas and Harlin call attention to studies in humid regions, including the long-term experience at Pennsylvania State University; the Muskegan, Michigan project with off season storage of treated wastewater, seasonal irrigation, and recovery of renovated water for surface discharge; the Belding, Michigan, study utilizing oxidation pond effluent for summer irrigation of forage, sod, and ornamentals and winter irrigation of forage on a sandy soil over a shallow water table which promotes lateral interflow with discharge to a surface stream; the Falmouth, Massachusetts, study and the Tallahassee, Florida, studies designed to demonstrate under radically different climatic conditions several management techniques on sandy soils with recharge to a sandy aquifer at 10 m.[8] Certainly this active area provides challenges to the many specialists who comprise the American Society of Agronomy for productive cooperation with water engineers and agencies dealing with water supply and wastewater treatment, disposal, or reuse.

Reuse of Agricultural Wastewater for Irrigation

For many years, dairies and other agricultural product processing plants have used their wastewaters to irrigate nearby fields. Discharge requirements arising from PL 92-500 now constrain such use. Here, it seems to me, is another challenge for agronomists to arrive at solutions which avoid water pollution but permit use of such water. Treatment and discharge requirements now being imposed on some processors threaten their continuance in business.

Runoff and wastewaters from animal feeding operations are now under controls, and much research is under way on proper waste and water management.

Return-flow waters from irrigation have been used directly or indirectly for years. Surface runoff as irrigation tailwater can be directed onto a lower-lying field or collected in a sump and pumped back into the water supply. Such water may have more or less sediment, phosphates, and pesticides, largely depending on sediment content, and very little more TDS than the initial supply water. Thomson and Allen (1975) call attention to phytophthora species in irrigation water collected in sumps for recycling to citrus orchards near Phoenix, Arizona. The incidence of species pathogenic to citrus was sufficient to be considered a distinct hazard.

Return-flow irrigation water percolating through the root zone and collected in drains will contain few nutrients or pesticides but higher salts. De-

[8] Thomas and Harlin, Jr., 1975. EPA research on land treatment.

pending on TDS, these waters are being used for irrigation but are discharged into drainage outlets when too saline. The Imperial Irrigation District provides each farmer with outlets to dispose of underdrain effluents from about 121,000 ha (300,000 acres) which drain into the Salton Sea. In some areas of the San Joaquin Valley, water tables are rising and salts accumulating, which will require a drainage outlet for disposal either in evaporation ponds, bird marshes, or in drainways ultimately leading to the ocean. Some environmental, technical, and institutional problems associated with disposal of wastewaters from the valley were discussed in an earlier section.

Uncollected return-flow water percolating below the root zone will either reach the groundwater table or be perched on slowly permeable materials. This water may be recovered and reused for irrigation or other purposes where suitable aquifers exist, the lifts are economic, and the water quality acceptable. This percolating water may be better or worse in quality than that in the basin, depending upon source of the irrigation water, irrigation efficiencies, soil characteristics, and other factors. As indicated by studies of Parker and associates under the NSF-RANN project, in some situations waters now percolating toward the water table but not yet arrived contain much higher N and TDS than the groundwater, indicating more serious problems ahead.[9] Where the irrigation water has been drawn from the same groundwater basin, the percolating water will be substantially higher in TDS than the pumped water because of the concentrating effect of transpiration. These considerations all cast doubt on the technical possibility of avoiding degradation of groundwater quality in basins which supply the irrigation water and receive little recharge by low-salt waters without importing low-salt water, resorting to desalting, or abandoning irrigation land (Santa Maria example, described in the section on Nonpoint Pollution of Groundwater Basins).

As previously mentioned, there are now suggestions that irrigators be required to reuse all wastewaters until they are no longer suitable for crop production. A lower limit on TDS might possibly be imposed before such return-flow water would be accepted in a drainage disposal system.

Water-Supply Exchanges Between Cities and Irrigated Farms

One of the National Water Commission's recommendations (1975, p. 128) is that arrangements for water exchanges should be considered in places where a high-quality source of water is now being used for irrigation of crops or golf courses and an unused source of treated municipal wastewater is available. Nutrient-rich municipal wastewater could be used beneficially for irrigation, leaving the high-quality water now used by irrigators for domestic and industrial uses that require a higher-quality water.

Such exchanges are under way in a few locations. One example is the City of Fresno, in California's San Joaquin Valley, which has an arrangement

[9]Kearney Foundation of Soil Sciences, 1973, 1974, 1975. Nitrate in effluents. Unpublished reports.

with the Fresno Irrigation District whereby pure mountain water contracted from the U. S. Bureau of Reclamation is made available to the city for recharge of the groundwater basin from which the city draws its supply, while the city's wastewater, after treatment and percolation through soil, is pumped and discharged into irrigation-district canals at a lower elevation. These arrangements, worked out after complicated negotiations aided by the agricultural knowledge of an extension soil and water scientist, appear to be benefiting both the city and the irrigation district, but some limitations remain in the use of this wastewater on some crops, such as sugar beets (*Beta vulgaris*) near maturation, because of undesirably high levels of nitrogen, and on other crops, becasue of restrictions imposed by public health.

Such exchanges, often desirable in theory and practice, will be difficult to arrange in many situations for technical, financial, institutional, and human resistance factors. The mayor of a small city surrounded by irrigated agriculture and facing an uncertain urban water supply and difficult to fulfill requirements on its waste discharges, asked the water agency supplying the irrigation district to help arrange a water exchange with the city. He was told that contractual arrangements would preclude such an arrangement. Understandably, farmers may also be reluctant to give up low-salt high-quality water, with no public health restrictions on its use, for treated municipal wastewater which, even though it may contain important amounts of essential nutrients N and P, will usually have higher TDS, carry public health-directed restrictions on its use, and may contain heavy metals or other constituents whose long-term effects on soil productivity and acceptability of crops remain uncertain.

In concluding this section, some comments in a recent article by McGowan (1975) are appropriate. Man can modify and relocate waste products. Choosing the modification and reuse most beneficial to industry, the public, and the environment is the objective of proper planning. Successful wastewater systems designed to achieve more than a single technical function are to be sought. There is more to be designed in a pollution-abatement system than pollution abatement. We must also be sensitive to such impacts as the direct and multiplier effects on local economies.

ENERGY REQUIREMENTS IN WATER SUPPLY, USE, AND WASTEWATER TREATMENT AND REUSE

Agencies, boards, commissions and other planners and decision-makers face extremely difficult problems in seeking to respond to today's societal attitudes and goals. Since widespread concern over water shortages and water pollution preceded the "energy crisis", recommendations have been made and are still being made by environmental groups, agencies, and by others calling for approaches which give little or no attention to energy requirements. As Congressman Harold Johnson in the opening discussion on Water and Energy at the National Conference on Water said, "the new economic situation involving energy has made many. . .planning decisions obsolete." Many of the decisions involving water management substantially affect agriculture and are being made without participation or even consultation with agronomists. This would appear to be an area where members of the American Society of Agronomy should be more active in bringing into the planning and decision process the expertise they have in water, soil, and plant behavior.

For Alternative Water Supplies

The dependence of irrigated agriculture on the large amount of energy required to develop, convey, and distribute irrigation water is now being recognized by responsible officials, especially in California where the major aqueduct systems require large amounts of power. The Director of the California State Department of Water Resources believes his agency may soon face a real crisis in obtaining sufficient energy at reasonable rates to pump water through its project facilities which supply water to agriculture for San Joaquin Valley, and urban areas around San Francisco Bay and southern California. He considers energy requirements to operate the state water project to be the single most serious problem facing the Department of Water Resources.

The energy supply problem now suggests that greater attention be given to the energy required to supply water from different sources.

An interesting illustration of the substantial differences in energy needed to supply a hectare-meter (acre-foot) of water from various sources to the Los Angeles Basin is shown in Table 3. Substantial differences also occur in energy required to supply water from different sources to agriculture in the San Joaquin Valley as shown in Table 4 using one water district. To meet the water needs in other service areas of California, the energy requirements of alternative methods and routes of supply deserve greater attention. A report on this and related water-energy relations has been completed (Roberts & Hagan, 1975) which will permit energy needs of alternatives to be considered in selection of the most suitable alternative.

For Irrigated Agriculture

A recent study has shown that in California during 1972, 68% of all electricity used by agriculture was used for pumping water and this amounted to 13.2% of all energy used by agriculture including manufacture of fertilizer, tillage, harvesting, and field processing of crops (V. Cervinka, W. J. Chancellor, R. J. Coffelt, R. G. Curley, and J. B. Dobie, 1974. Energy requirements for agriculture in California. Joint study, California Department of Food and Agriculture and University of California, Davis). The energy consumption reported apparently does not include the substantial energy required where water is supplied to the water district through major aqueduct systems requiring extensive pumping.

Table 3. Energy required to supply water to the Los Angeles Basin.

Source	Energy	Elevation
	KWH/ha-m	m
Los Angeles aqueducts	–19,464*	518
Groundwater pumping	3,244	varies
Colorado River aqueduct	16,828	457
State Water Project (avg.)	23,519	518
Wastewater		
Tertiary treatment†	11,962	0
Tertiary plus electrodialysis†	22,911	0
Seawater distillation	170,310	0

* Power generated

† Added energy to treat beyond level of purity now required for discharge to ocean.

Table 4. Energy required to supply water to agriculture in the Carwelo Water District of the San Joaquin Valley

Cawelo Water District	Energy
	KWH/ha-m
Groundwater pumping (1971)	6,245
Groundwater pumping (after 100 years)†	11,111
East Side Project	9,448
Enlarged Cross Valley Canal	4,850
State Water Project	3,917

* Pumping power only.

† Assumes 1 m/year drop in water table.

Interesting studies by Chancellor indicate that 45% of all the energy needed to produce California crops and to deliver them to the processors is required to supply and apply irrigation water (W. J. Chancellor, 1975. Unpublished studies of Department of Agricultural Engineering, University of California, Davis). This is direct energy consumption and does not include energy to manufacture or maintain the pumps. In this production budget, 22% of the remaining energy goes for fertilizer (manufacture, transport, and application) and 34% for cultivation, pesticide applications, and harvesting (again not including energy for manufacture, etc.). Chancellor and Singh (1975) report the new high-yielding wheats grown in northern India, have a direct energy input for irrigation of 45% and fertilizer 18%; for sugarcane (*Saccharium officinarium*), one year crop, irrigation 53% and fertilizer 23%; and for maize, grown during the monsoon with only supplemental irrigation, 25% for irrigation and 32% for fertilizer. Chancellor and Singh conclude that, under reasonably intensive irrigated agriculture, energy for irrigation takes nearly half of the total energy used for crop production and delivery wherever irrigation supplies most of the water used. In the midwest U. S., obviously, irrigation energy requirements would be zero and would increase as supplemental irrigation is employed.

A study is being initiated with the Energy Research and Development Administration to determine the full energy requirements for supplying and applying irrigation water to the Central Valley of California—as influenced by source of irrigation water, irrigation method and its operation, climate in various regions, soils and other site conditions, and types of crops grown (Department of Agricultural Engineering and of Water Science and Engineering, University of California, Davis, 1974. Energy required to irrigate crops in the California Central Valley and Delta area. Contract No. W-7405-Eng-48). Information obtained will be used to evaluate the total effects on energy consumption of changes in location of irrigated lands, land preparation including grading, irrigation methods, irrigation system design and operation, irrigation scheduling, source of irrigation water, off-peak and on-peak pumping, such practices as tailwater return systems, cropping patterns, and other related factors.

It is clear that rising energy costs must be reflected in higher water costs. Conceivably this could bring about changes in cropping patterns and even relocation of some irrigated agriculture.

For Different Irrigation Systems

The following examples of energy consumption by different irrigation systems illustrate that where the energy required to deliver water to the field is low, replacement of surface irrigation by a more efficient sprinkler irrigation system can result in an increased energy consumption to meet the seasonal irrigation requirement, but where higher amounts of energy are needed to deliver water to the farmer, the more efficient sprinkler system, by reducing the amount of water required, can reduce the seasonal energy requirement despite its higher energy input per unit of water applied (Table 5).

Table 5. Energy consumption for irrigation as affected by energy to supply water and type of irrigation system

Energy requirement for	Surface irrigation		Sprinkler irrigation	
		KWH/ha		KWH/ha
(1) Low energy requirement for water supply (811 KWH/ha-m)				
Water supply	0.609 ha-m/ha X 811 KWH/ha-m =	494	0.396 ha-m/ha X 811 KWH/ha-m =	321
Application	(gravity)	0	0.396 ha-m/ha X 1,135 KWH/ha-m =	449
Totals		494		770
(2) Higher energy requirement for water supply (16,220 KWH/ha-m)				
Water supply	0.609 ha-m/ha X 16,220 KWH/ha-m =	9,880	0.396 ha-m/ha X 16,220 KWH/ha-m =	6,420
Application	(gravity)	0	0.396 ha-m/ha X 1,135 KWH/ha-m =	449
Totals		9,880		6,869

A recent study by Kizer et al. (1975) develops an energy consumptive model (Fig. 7) to predict the total energy required to manufacture, install, transport, and operate pressurized irrigation systems including hand-move, side-roll, center-pivot, solid-set, permanent and drip (M. A. Kizer, R. B. Wensink, J. W. Wolfe, M. N. Shearer, 1975. Energy model for pressurized irrigation systems. Am. Soc. of Civ. Eng., Irrigation and Drainage Specialty Conference, Logan, Utah). The model analyzes specific irrigation systems by

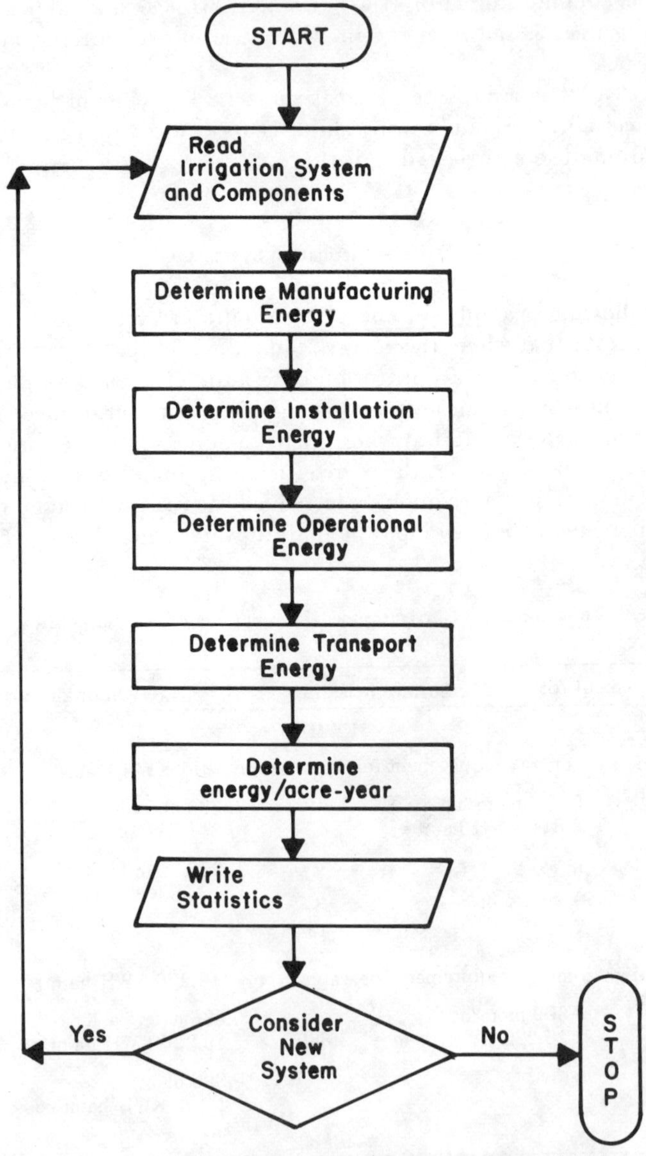

Figure 7. A flowchart of irrigation energy utilization model.

modularizing the energy requirements within each system into four subsystems—manufacturing, installation, operation, and transport. The irrigation season energy per hectare (acre) requirements ranged from a high of 17,323 KWH/ha-m (2,136 KWH/acre-foot) for a permanent over-tree-system to 4,671 KWH/ha-m (576 KWH/acre-foot) for a drip system on the same 8 hectare (20 acre) orchard in Oregon.

For Water Quality Control in Irrigation Return Flows

Environmental concerns are leading to numerous proposals for water quality control that have substantial energy requirements. These are still generally overlooked by the promoting organization and by the general public. Expected future energy shortages point to the need for greater attention to this aspect in seeking to conserve water and to implement the provisions of PL 92-500.

Concerned environmental groups are bringing pressure (including lawsuits) on EPA to require rapid compliance with PL 92-500. Yet some goals of this Act still need clarification, the technology required remains uncertain, and energy requirements of alternative solutions are still largely unexplored. Some approaches to achieving water quality improvements, involving substantial energy inputs, are now being vigorously advocated and pressed in litigation at local, state, and federal levels.

An example of this present situation is provided by the decision by the U. S. Government to build a large desalting plant to process irrigation return flows from the Mohawk-Wellton Project in Arizona and thus improve the quality of the Colorado River as it flows into Mexico. This decision commits our nation to large annual outlays of energy to achieve water quality improvements which may have been possible, in part at least, by using other less energy-intensive approaches. Complex political, economic, and institutional factors, and especially pressures from special interest groups and a concerned public, may cause similar decisions to be made before the energy consequences of alternative measures can be adequately evaluated and rational trade-offs arranged. Unfortunately actions will likely be taken and public policies set within the next few years which will be essentially irreversible or capable of modification only at great cost.

In the technical evaluations of water quality control measures made by public and private engineering agencies, energy requirements are generally reported only as the overall dollar cost for purchased power. To date, it appears that little has been done to collect, retain, and disseminate comparative data. As a result, scanty information is available, either in published or unpublished form, upon which to base some computations of the energy requirements for water quality control measures.

It is clear that information on energy requirements of alternative approaches to water quality management is badly needed by environmental groups, planners, and especially decision-makers at all levels.

For Reuse of Municipal Wastewater for Irrigation

Renovation and reuse of urban wastewater for irrigation is increasingly advocated to conserve water and to provide an alternative to tertiary treatment of effluents before discharge to receiving waters. Little attention has been given to energy requirements of conveying treated effluent and applying it to areas suitable for irrigation. Studies should be made of possible energy savings by substituting land treatment of secondary treated effluent for tertiary treatment. Such energy savings would reduce the energy requirement in using urban effuent for irrigation.

Another aspect to be considered in analyzing the total energy balance in reusing wastewater is the energy value of contained nutrients. Considerable energy is required for the manufacture of nutrients and, if they can be efficiently used by crops to which the wastewater is applied, they do represent an energy saving appropriate to consider (Table 6). It should also be recognized these nutrients, when present at certain stages of crop growth (such as maturation of sugar beets), may actually reduce yields.

SUMMARY

Water management is a complex problem. The general public and, in some cases, even the personnel of planning and regulatory agencies, do not realize how many intricate pieces must be fitted together if a properly integrated and coherent water management program is to be achieved which will properly deal with our whole ecosystem.

In today's complex world it is easy to solve one problem while unwittingly creating many other problems, some more widespread and difficult to resolve than the original problem. So, in water management, including efforts at water conservation and pollution control, society may be creating new problems–technical, economic, social, and legal. For example, a difficult obstacle to programs of pollution control (and also water conservation) is the institutional structure of western water laws. Any "water savings" may diminish water rights if less water is diverted over time.

It is important that all of us work actively to bridge the serious communication gap, and consequent lack of understanding, between urban and rural dwellers and between environmentalists, agencies, water and soil scientists and other agronomists, economists, social scientists, and others in the general public who now hold divergent views–aggravated by emotions and widely different philosophies about life and the social system. The resultant polarization is hindering progress toward solution of many water management problems.

Salinity as a pollutant will remain the most serious problem in western states. Salts cannot be legislated away at either state or federal levels. If pollutants, especially salts, are to be eliminated, or at least reduced, from our lakes, streams, and groundwaters, then these pollutants must be transferred elsewhere. But where on our spaceship planet?

Table 6. Energy value of nutrients in wastewater

Plant nutrient	Content in effluent	Energy to produce, transport & apply	Energy value of nutrients
	kg/ha-m	KWH/kg	KWH/ha-m
N	250	6.15	1,540
P_2O_5	460	0.225	105
K_2O	360	0.225	81
Total			1,726

Obviously, the water pollution problems are much different in the humid states and vary also depending upon the primary sources of pollutants —municipal, industrial, or agricultural. Fish, other aquatics and wildlife, and recreation will generally benefit from depolluted waters, but how does the pollutant transfer process affect the rest of the ecosystem, particularly the terrestrial aspects? These fundamental questions are raised in an unpublished report from the Utah Water Research Laboratory (personal communication). That report points out that environmental degradation means more than simply pollution of water and, likewise, that enhanced environmental quality should consider more than simply an "improved" aquatic environment. Certainly some nutrients are essential in water to maintain aquatic life, and added nutrients in some waters may produce a more desirable productivity.

Finally a question which needs to be seriously addressed: Is our whole ecosystem best protected by attempting to retain the common salts—produced by natural weathering and concentrated by plant transpiration losses—in our soils, groundwaters, or newly established evaporation ponds, or should provisions be made to convey waters, when too salty for further use, to the planet's great salt sink, the ocean?

A related policy question: Is this total ecosystem best preserved by increased use of soil for wastewater treatment and of wastewaters for irrigation? This is an especially difficult question in arid states with major salt problems.

A second basic question involves energy. Often the saving of water will save energy, but some measures proposed to conserve water will increase power requirements. Where water conservation, pollution control, or wastewater treatment and reuse increase energy needs, serious attention needs to be directed to a basic water and resource management question: Is it sound public policy to consume energy (often a nonrenewable resource) to conserve water (a naturally renewable resource through the operation of nature's great desalting plant, the hydrologic cycle)?

Water management decisions, on-farm and at water agency levels, as well as many other farm, industrial, and domestic activities, will be determined by society's answers to these and other equally basic questions.

This whole area of water and the environment is full of tremendous challenges to agronomists, other professionals, and to clear-thinking opinion-shapers and decision-makers.

LITERATURE CITED

Ayers, R. S., and R. L. Branson, eds. 1973. Nitrates in upper Santa Ana River basin in relation to groundwater pollution. Calif. Agric. Exp. Stn. Bull. 861.

Ayers, R. S., and R. Coppock. 1974. Salt management: California's most complex water quality problem. Environmental Concerns Leaflet, University of California Cooperative Ext. Serv.

California Department of Water Resources. 1976. Water Conservation in California. Bulletin No. 198.

Chancellor, W. J., and G. Singh. 1975. Energy inputs and agricultural production under various regimes of mechanization in northern India. Am. Soc. Agric. Eng., Trans. 18(2):252–259.

Coppock, R., and V. P. Osterli, eds. 1975. Nitrogen: food production and water quality. California's Environment, University of California Cooperative Ext. Serv. No. 25, p. 3–5.

Dreyfus, D. A. 1975. 2020 hindsight: another fifty years of irrigation. Am. Soc. Civ. Eng. Proc., J. Irrig. Drain. Div. 101(IR2):87–94.

Environmental Protection Agency. 1973. The mineral quality problem in the Colorado River Basin. Summary Report and Appendix A Natural and man-made conditions affecting mineral quality, Appendix B Physical and economic impacts, and Appendix C Salinity control and management aspects. Washington, D. C.

Environmental Protection Agency. 1975. Evaluation of land application systems. Tech. Bull. EPA-430/9-75-001. Washington, D. C.

Hagan, R. M., H. R. Haise, and T. W. Edminster, ed. 1967a. Irrigation of agricultural lands. Agronomy 11. Am. Soc. of Agron., Madison, Wis.

Hagan, R. M., C. E. Houston, and R. H. Burgy. 1967b. More crop per drop: approaches to increasing production from limited water resources. Int. Conf. on Water for Peace, Wash., D. C., May 23–31, 1967. Document No. P/551.

Hanks, R. J. 1974. Model for predicting plant yield as influenced by water use. Agron. J. 66:660–665.

Hanks, R. J., J. Keller, and J. W. Bauder. 1974. Line source sprinkler plot irrigation for continuous variable water and fertilizer studies on small areas. CUSUSWASH 211D-7, Utah State University.

Hsiao, T. C. 1973. Plant responses to water stress. Annu. Rev. Plant Physiol. 24:519–570.

Jensen, M. E. 1969. Scheduling irrigations with computers. J. Soil Water Conserv. 24: 193–195.

Jensen, M. E., and H. R. Haise. 1963. Estimating evapotranspiration from solar radiation. Am. Soc. Civ. Eng. Proc., J. Irrig. Drain. Div. 89 (IR4):15–41.

Jensen, M. E., D. C. N. Robb, and C. E. Franzoy. 1970. Scheduling irrigations using climate-crop-soil data. Am. Soc. Civ. Eng. Proc., J. Irrig. Drain. Div. 96 (IRI):25–38.

Kleinman, A. P., G. J. Barney, and S. G. Titmus. 1974. Economic impacts of changes in salinity levels of the Colorado River. U. S. Bureau of Reclamation, Denver, Colorado.

Krantz, B. A. 1975. ICRISAT farming systems annual report, 1974–75. Int. Crops Res. Inst. for the Semiarid Tropics. Hyderabad, India.

McGowan, F. M. 1975. Water and land oriented wastewater treatment system. *In* Water Pollution Control in Low Density Areas. Proc. of a Rural Eng. Conf., publ. for University of Vermont by University Press of New England, Hanover, New Hampshire.

National Academy of Sciences–National Academy of Engineering. 1972. Water quality criteria. U. S. Government Printing Office, Washington, D. C.

National Academy of Sciences–National Research Council. 1974. More water for arid lands–promising technologies and research opportunities. U. S. Government Printing Office, Washington, D. C.

National Water Commission. 1973. Water policies for the future. Final report. Washington, D. C.

Rawlins, S. L., and P. A. C. Raats. 1975. Prospects for high frequency irrigation. Science 188:604–610.

Roberts, E. B., and R. M. Hagan. 1975. Energy requirements of alternatives in water supply, use, and conservation: a preliminary report. University of California Water Resources Center. Contribution No. 155.

Robinson, F. E. 1974. Salinity management options for the Colorado River. Report from Office of Water Research and Technology Project No. B-107-Utah and University of California Water Resources Center Project UCAL-WRC-W-482.

Second International Drip Irrigation Congress. 1974. Proceedings. San Diego, California, July 7-14.

Slatyer, R. O. 1969. Physiological significance of internal water relations. p. 53-88. *In* J. D. Eastin, F. A. Haskins, C. Y. Sullivan, and C. H. M. van Bavel (ed.) Physiological aspects of crop yield. Am. Soc. of Agron., Madison, Wis.

Thomson, S. V., and R. M. Allen. 1975. Phytophthora species in Arizona: Its occurrence in recycled irrigation water. Prog. Agric. Agriz. 27(3):8-9.

U. S. Army Corps of Engineers. 1972. Wastewater management by disposal on the land. Spec. Rep. No. 171, Cold Regions Res. and Eng. Lab., New Hampshire.

U. S. Bureau of Reclamation. 1974a. An appraisal of total water management in the Central Valley Basin, California. Mid-Pacific Region, Sacramento, California. August, 1972. Revised January, 1974.

U. S. Bureau of Reclamation. 1974b. Colorado River quality improvement program: status report, USBR, Denver, Colorado.

Valantine, V. E. 1974. Impacts of Colorado River salinity. Am. Soc. Civ. Eng. Proc., J. Irrig. Drain. Div. 100(IR4):495-510.

van Schilfgaarde, J., L. Bernstein, J. D. Rhoades, and S. L. Rawlins. 1974. Irrigation management for salt control. Am. Soc. Civ. Eng. Proc., J. Irrig. Drain. Div. 100(IR3):321-338.

Improving Germplasm Resources

Louis P. Reitz

Germplasm, the grist of every plant breeding program and a basic component of every well-planned biological study, is prominent among current topics (Creech & Reitz, 1971; Frankel & Bennett, 1970; Harlan, 1975; National Academy of Science, 1972) but not a new subject (Hodgson, 1961). This is encouraging and is a precondition to any improvement in dealing with this tremendous natural resource. Concern and then action must mount until our zeal for saving so-called endangered species is surpassed by our zeal for safeguarding our germplasm. Indeed, man will put his food supply in jeopardy and may well end up on the endangered species list, if he is not already there, should he neglect to protect the basic germplasm resource nature has provided. The American Society of Agronomy should make no apology for a strong bias in this regard.

Crop germplasm has been defined (Creech & Reitz, 1971) as an array of plant materials, assembled or not, that serves as a basis for crop improvement or related research. Its chief characteristic is that of a reservoir of genes available to meet the needs of plant breeders both now and tomorrow. Included in the definition, and interpreted more narrowly, is the term genetic stocks. The Crop Science Society of America restricts this concept to stocks of specific genes and gene combinations that have direct usefulness in genetic analyses. We should regard the two kinds of stocks as complementary, differing mainly in use, specific requirements for maintenance, and kind of documentation of merit.

Examples of benefit from germplasm are legion. A few among cereal grains are the following: In 1874, the U. S. agricultural attache to Japan described in his reports the Daruma type of short-straw wheat (*Triticum aestivum* L.) utilized in that country (Reitz, 1970). No use whatever was made of this information. In 1946, 72 years later, S. C. Salmon, U. S. Department of Agriculture (USDA) agronomist, rediscovered the Daruma types there, collected 16 of them including 'Norin 10', and made them available to breeders in the U. S. From this introduction, the first successful semi-dwarfs were bred in America.

Opaque-2 was the name given in 1938 to a genetic type of corn (*Zea mays* L.) with defective endosperm. It was discovered some 30 years later to

L. P. Reitz is cereal specialist, National Program Staff, U. S. Department of Agriculture, retired.

have an altered ratio of amino acids in the protein and the relative content of lysine was much higher than normal.

Hiproly barley (*Hordeum vulgare* L.) (C. I. 3947) and a sister line (C. I. 4362) were introduced from Ethiopia and made a part of the USDA world collection in 1924. In 1970, 46 years later, high protein and high lysine contents were discovered in these selections.

Long before the cereal leaf beetle reached Michigan (about 1962), a number of resistant genotypes of wheat were already in the USDA collection. One of these (C. I. 9321) entered the collection from Saratov, Russia in 1927. Its hairy leaves provide a mechanism of resistance to three stages of the beetle. No one knew it was there for 40 years, and no one knew why such miserable looking material was being kept.

In 1874, Turkey wheat was introduced by immigrants. It was the cornerstone of our hard red winter wheat industry (Quisenberry & Reitz, 1974) and, because it contained many different genotypes, was reselected to provide numerous cultivars. These, in turn, were used widely in cross-breeding.

Triticum timopheevi became a part of the USDA collection in 1930 (P. I. 94760). There, 32 years later, it was discovered to have the mechanism now being used to produce hybrid wheat. D. Brezhnev, Director of the Vavilov All-Union Institute of Plant Industry, Leningrad, U.S.S.R. told me that recent collecting efforts in the Caucausus where *timopheevi* was found in 1930 did not reveal it. Presumably, it has disappeared from the wild state.

In each of the preceding examples, more than one generation of breeders passed from the time the germplasm came into a collection until it was used. Of course, there are many stocks that are utilized almost at once. The core of our concern is that useful stocks be available when the need arises. Where will we get germplasm in the future?

The greatest array of plant germplasm, in general, has existed in farmer's fields, in wild populations on waste, cultivated, forest, swamp, seashore, and other lands, and in the ocean (Fig. 1). But plants, no less than animals, become extinct both from natural causes and because of the activities of man. Certainly man is an accelerating element in this process since his utilization of the earth's surface increases as his numbers increase. He exercises depletive harvesting practices of cutting, stripping, and grazing, and he denudes the land to build roads and cities, and engages in numerous other plant-destroying activities. He also uses the land to dump all kinds of refuse, which spoils both the immediate site and the ground, water, and air surrounding it.

CONCERN FOR GERMPLASM

In an arresting discussion of large animal extinctions in America, Martin (1975) showed that big game hunters and the (now) extinct animals coexisted for no more than 10 years. His models and data showed that after 350 years of migration from Alaska and Canada to the Gulf of Mexico, the relatively

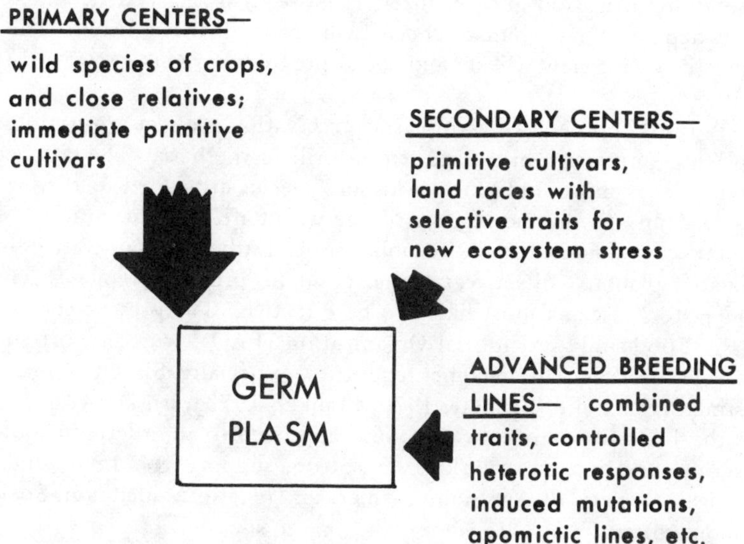

Figure 1. Sources of germplasm relative to amount and diversity of contributions (Creech & Reitz, 1971).

sparse human population with its rather wasteful harvesting practices exterminated some 29 species of large mammals. This rate of decimation was repeated on several animals, birds, and plants during the expansion of settlement westward from the eastern states in the period 1775 to 1875. Conservation was almost too late to save from extinction a host of species now on the endangered list.

Our concern for germplasm should extend into all of these areas. Specifically, the American Society of Agronomy should give scientific leadership to this concern. Yes, national forests, wilderness areas, and protected preserves do represent approaches to responsibility, but these are relatively small areas, and they come under constant pressure for other uses that are depletive.

The next largest array of germplasm, and the main source of crop germplasm, has been and continues to be farm fields, pastures, managed woodlots, gardens, orchards, and protected range. Also city parks, arboretums, cemeteries, and other such sites contain a huge array of germplasm though most of it is of nonfood plants. Unfortunately for germplasm, the array in farm fields has been shrinking at an accelerating pace the last 25 years or so (Frankel & Bennett, 1970; Harlan, 1972; National Academy of Science, 1972). In fact, rapid replacement of land races of most crops with a narrower range of improved cultivars is the reason for so much recent agitation about the need for preservation of germplasm.

Land races of crops fill not only biological niches but also technological and social niches. Hence, a change in any of these niches may spell doom for the germplasm associated with it. For example, a superior new cultivar that responds to irrigation and fertilizer, is more easily harvested, and that reduced dependence on human energy will erode all three niches. The interactions involving new wheats and social progress have been reviewed (Reitz, 1970).

A third approach to germplasm preservation is direct intervention, that is, placing germplasm in storage centers or living herbaria. The task becomes formidable as more and more crops and species are added to the list to be preserved and as diverse regions of the world are given attention. No one country can assume the entire obligation. Rather, a cooperative and coordinated plan to collect, catalogue, evaluate, preserve, replenish, and distribute stocks seems most likely to be effective. The present coordination by the Food and Agricultural Organization (FAO), with support and considerable prodding by member countries, especially Sir Otto Frankel of Australia (Frankel & Bennett, 1970; Frankel, 1973a, b), and countries outside the FAO represents the first joint approach to an orderly plan of germplasm preservation on a world scale. Strong support from the Foundations, American Society of Agronomy, American Genetic Association, Society of Economic Botany, etc. are helping the plan to succeed.

World interest in germplasm collections and storage has grown at an accelerating rate in recent years (FAO, 1970, 1972, 1973; Frankel & Bennett, 1970). Global coordination is being fostered and offered by FAO and the International Biological Program (Frankel, 1973b). Through the latter there became available lists of curators of many crops little known about before (FAO, 1970, 1972, 1973). A partial listing of major world centers appears in Table 1.

Last year I had the opportunity to visit both the U.S.S.R. and People's Republic of China. Perhaps some observations will be of interest. The U.S.S.R. show great aggressiveness in the conservation of germplasm. They are interested in new exploration of their own flora and the flora of key areas of the world. They have rejuvenated their germplasm program and have just completed a seed storage center that is as large, they say larger, than the one at Fort Collins, Colorado. The concepts and objectives of N. I. Vavilov, Russia's most illustrious plant explorer, are highly praised. The Vavilov Institute of Plant Industry (VIR) at Leningrad is well supported, and evaluation activities are carried on at some 20 stations with added cooperation at most of the breeding centers all over the country. There is an eagerness and sense of mission displayed by the employees of VIR. My visits to nine institutes where cereal evaluation work was being done impressed me with the scope and genuineness of their effort (Anon, 1973).

My observations in the People's Republic of China were limited by my mission there, but my impression corresponds to that of the Plant Studies Delegation which visited the People's Republic of China last year (NAS, 1975). China has a diverse agriculture with thousands of local, usually small units, on which separate judgments may be reached about the best cultivars to multiply. This situation tends to delay erosion of land races. However, seed production brigades are now taking over more of this determination and little attention is given to germplasm collecting. As a result, China's lack of

Table 1. Some of the major field crop germplasm storage centers in the world*

Center and location	Crop seeds stored†
National Seed Storage Lab., USDA, Ft. Collins, Colorado	all kinds
N. I. Vavilov All-Union Institute of Plant Industry, Leningrad, U.S.S.R.	all kinds
Crop Research and Introduction Center, Ismir, Turkey	cereals, grain legumes
International Rice Research Institute, Los Banos, Philippines	rice
International Crops Research Institute for the Semiarid Tropics (ICRISAT), Hyderabad, India	sorghum, peanuts, millet, chickpeas, pigeonpeas
Laboratorio del Germoplasma, Inst. di Agronomia, Bari, Italy	cereals, vetch, pea, etc.
CIMMYT, El Batan, Mexico	corn
Instituto Colombiano Agropecuario, Colombia	corn
Centro Internacional de Agricultura Tropical, Cali, Colombia	cassava, beans
International Inst. of Tropical Agric., Nigeria	cowpeas, yams, pigeonpeas, tubers
National Seed Storage Facility, Hiratsuka, Japan	rice, wheat
INIA, Colegio de Postgraduados, Chapingo, Mexico	beans, peppers
World Collection of Wheat, Tamworth, Australia	wheat
Waite Agric. Res. Institute, Adelaide, Australia	barley
Central Exp. Farm, Ottawa, Ontario, Canada	oats, wheat, barley
Ohara Inst. for Agric. Biol., Okayama, Univ. Kurashiki, Japan	barley
Foundation for Agric. Plant Breeding, Wageningen, Netherlands	barley, oats, wheat
Swedish Seed Assoc., Svalof, Sweden	cereals

* For more comprehensive listings see FAO (1970, 1972, 1973) and Frankel (1973b).
† The botanical names for the crops are listed in alphabetical order as follows:

barley (*Hordeum* spp.)
beans (*Phaseolus* spp.)
cassava (*Manihot* spp.)
chickpeas (*Cicer* spp.)
corn (*Zea mays* L.)
cowpeas (*Vigna* spp.)
oats (*Avena* spp.)
pea (*Pisum* spp.)
peanuts (*Arachis* spp.)
peppers (*Capsicum* spp.)
pigeonpeas (*Cajanus* spp.)
rice (*Oryza* spp.)
sorghum (*Sorghum* spp.)
vetch (*Vicia* spp.)
wheat (*Triticum* spp.)
yams (*Dioscorea* spp.)

concern for diverse germplasm is placing this rich resource in jeopardy though the process is more delayed than in some areas. It is my hope that the People's Republic of China will respond to the need of China and that of the world and help make germplasm more widely available.

COLLECTING GERMPLASM

One who has not done some phase of germplasm maintenance can not grasp the size of the undertaking in time, facilities, and funds. For example, the USDA small grain collection of over 70,000 items at Beltsville occupies the entire time of one professional agronomist, three technicians, and some part-time labor; 293 m^2 (3,150 ft^2) of floor space for the seed; and 1.6 ha (4 acres) of land for growing materials. The costs are not flexible and must be met every year, forever. Standby stocks duplicating all those held at Beltsville are kept in storage at the National Seed Storage Laboratory (NSSL) at Fort Collins, Colorado. There are additional expenditures of effort, funds, and facilities made by the collectors and evaluators of germplasm.

Such collections are vulnerable to accidental destruction, loss from lack of technical attention, the withdrawal of commitment by the sponsoring agency, government instability, the vagaries of war, and complacency. Binational agreements, such as the U.S.S.R.–U.S.A. agricultural agreement, have included germplasm collection, maintenance and evaluation among high priority items. Under the agreement with Russia, we expect to exchange listings of the main cereal collections each holds, beginning with wheat. The exchange of 20,000 wheats will be a first step in extending the collections and in introducing greater safety in the preservation of these stocks. Fortunately curators from a dozen other countries are being invited to participate in the wheat program, including the People's Republic of China.

Following the publication of the report "Genetic Vulnerability of Major Crops" (NAS, 1972), a USDA, Agricultural Experiment Stations (AES), and industry task force undertook to outline what ought to be done about it. The result was "Recommended Actions and Policies for Minimizing the Genetic Vulnerability of Our Major Crops" (USDA & National Association of State Universities and Land Grant Colleges, 1973).

The task force recommended that a National Plant Genetics Resources Board be established, that a broad plan for plant germplasm resources be developed, and that we greatly expand research and other efforts to conserve and manage our own and the world's plant germplasm. The Secretary of Agriculture has authorized action on the first recommendation. He expects the Board to provide leadership and counsel on actions and policies regarding the collection, maintenance, and utilization of plant germplasm; to coordinate collection plans among several agencies including international ones; and to assess the national needs and identify high priority programs for conserving and utilizing plant genetic materials to minimize genetic vulnerability (Wright, 1975).

Simultaneously, the USDA and AES looked at their programs and sought a more meaningful way to coordinate their work. There now exists a clearer definition of responsibility and awareness than I have ever known. A chart showing how these groups and the new Board may work is given in Figure 2. Breeders of specific crops, germplasm curators, storage centers, and sponsors of research and evaluation all have a place or are given recognition.

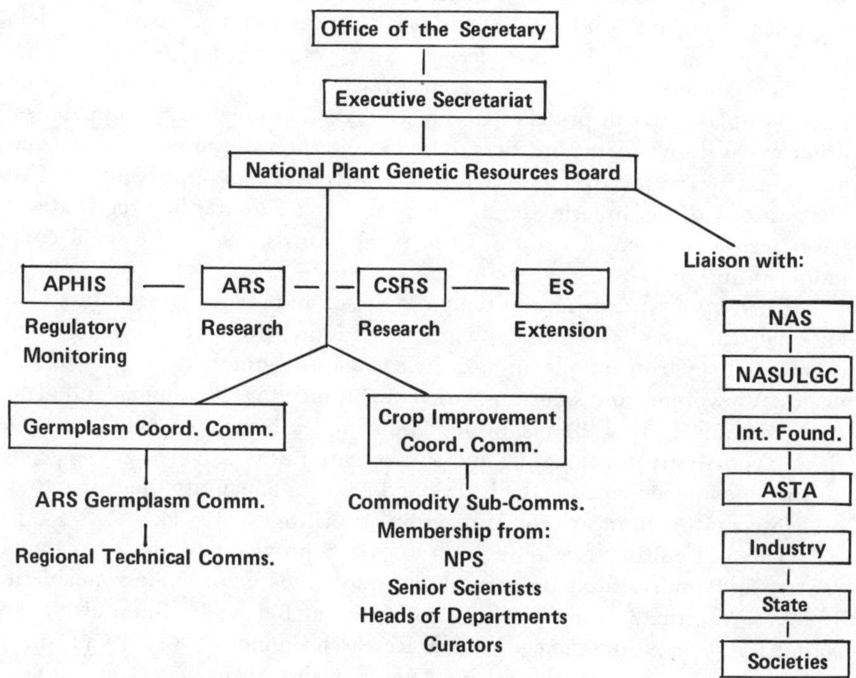

Figure 2. The Plant Genetics Resources Board can provide for coordination of the work of many agencies and worldwide cooperation.

GERMPLASM PRESERVATION

Budget makers groan at the cost of germplasm maintenance and rightly so, because it does cost a lot. Every commodity group should be challenged anew, and repeatedly, to develop and implement more efficient ways to preserve germplasm. This means research must be done on ways to lengthen the life of seeds, ways to assess the numbers needed to represent a species, ways to intelligently composite and extract genotypes, and ways to save nonseed-forming species by tissue culture of some kind. For example, liquid nitrogen storage has enhanced genetic work with fungi and other microorganisms and may have wider use. Automated data processing and mechanical seed packeting are already making these tedious processes easier, quicker, and cheaper.

Every plant breeding and genetics unit has a collection of germplasm that I suspect is eventually wasted because so little is kept. Only a few breeders keep material longer than a few years. One Agronomy Department Head told me that his staff had on hand at all times between 100,000 and 125,000 items in addition to the thousands of lines in their segregating populations. I suspect this is a common situation at major breeding centers. Of course, much of this material is of passing interest, but considerable amounts have value for conservation and should not be overlooked.

One breeder told me of materials with outstanding potential for disease control, but he was too busy to send it in for safe-keeping. He has since left the project, and I suspect his successor "threw that old junk out." Just because a strain does not become a commercial cultivar is no reason to ignore it as germplasm worth preserving. The CSSA plan to register germplasm is a prime motivation to some breeders to be more responsible about unique combinations, new mutations, and useful composites. In fact approximately 340 registrations of germplasm for 22 crops and 132 parental line registrations have been published in *Crop Science* (C. F. Lewis, 1975. Personal communication).

I had my first exposure to the problems of germplasm preservation on a national scale over 30 years ago when the National Research Council called representatives from public and private agencies together to assess status and needs. Viewpoints and solutions varied at this meeting, but animal and plant scientists quickly saw the magnitude and complexity of the problem in each field. Apparently this meeting and subsequent study provided guidance, or at least encouragement, for the USDA and AES to inaugurate their first comprehensive effort in regard to plant germplasm (Burgess, 1971).

The Agriculture Marketing Act of 1946 provided funds for regional centers and for research on germplasm, and funds were made available in 1948 for the small grains working collection at Beltsville. Soon after, the National Academy of Science-National Research Council (NAS-NRC) formed a committee on preservation of indigenous strains of maize. Over a 3-year period, through a broad cooperative effort, 15,000 variants of maize were collected, evaluated, grouped into races, and made available to breeders. This collection has had its bad times, but in the main it has been retained. A partial tabulation of germplasm in major working stocks is given for the U. S. in Table 2. Most of this seed is in or destined to go to the NSSL at Fort Collins, Colorado for long-term storage. The work is impressive (Burgess, 1971; Creech & Reitz, 1971; Hodgson, 1961; USDA-ARS, 1974).

At our meeting 30 years ago, the question about adequate numbers that should be preserved came up for discussion. It always does. We are still discussing it. When additions are made to even a small collection, repetition of genetic traits becomes apparent, and this repetition is all the more apparent in large collections. Do we need 35,000 wheat or 12,000 maize items to represent these species? The answer I usually hear to this question is "No, not if you have the right items included." But agreement cannot be reached with confidence on which items are the right ones. Certainly, at a minimum, you would maintain the separate species, or subspecies if you are a species splitter. Breeders are never satisfied with a collection unless they find in it the characteristic they want. This leaves the number an open question.

Our objective can be served by wide-scale sampling the primary and secondary centers of diversity the world over. Concentration on the primitive forms seems most urgent at this time. Such a reservoir would provide breeders with material for further development by selection and crossbreeding or by subjecting it to mutagens, but it would be a slow process to

Table 2. A sample listing of some of the major working stocks of crop germplasm in the U. S.*

Center and location	Crop†‡	No. items
Agric. Res. Center, PGGI, Beltsville, MD	wheat (20)	35,000
	barley (2)	18,000
	oats (7)	14,000
	rye & triticale (3)	3,000
	rice (2)	10,000
	tobacco (1)	1,000
Interregional Project IR-1, Sturgeon Bay, WI	potato (82)	3,000
Northeast Project NE-9 and S-9, Geneva, NY	clover (84)	1,835
Northeast Project NE-9 and W-6, Geneva, NY	onion (±40)	544
Northeast Project NE-9, Geneva, NY	pea (3)	1,708
Southern Project S-9, Experiment, GA	cantaloupe (20)	1,729
	cowpea (19)	1,623
	millet (3)	367
	peanut (16)	3,776
	pepper (8)	1,728
	sorghum (40)	3,423
Western Project W-6, Pullman, WA	beans (26)	6,034
	cabbage (1)	626
	fescue (53)	794
	lentils (4)	508
	lettuce (1)	476
	orchardgrass (11)	738
	ryegrass (17)	466
	safflower (16)	1,289
	sunflower	1,300
Northcentral Project, Ames, IA	alfalfa (13)	876
	bromegrass	548
	corn (1)	2,418
	cucumber (1)	537
	sugar beet (1)	300
	sweet clover (20)	525
	tomato (6)	3,500
	winter squash	250
	sunflower	600
Tobacco Res. Stn., USDA, Oxford, NC	tobacco (63)	490
Federal Exp. Stn., USDA, Mayaguez, P.R.	yams (8)	388
Alisal Branch Stn., USDA, Salinas, CA	sugar beet (6)	200
Agric. Exp. Stn., Raleigh, NC	tobacco (1)	±200
Agric. Exp. Stn., College Sta., TX	cotton (32)	936
	sorghum (3)	16,000
U. S. Regional Soybean Lab., USDA, Urbana, IL	soybean (1)	4,000
Agric. Exp. Stn., Urbana, IL	corn mutant (1)	500
Agric. Exp. Stn., Ft. Collins, CO	barley genetic stock (1)	2,000
Agric. Exp. Stn., Columbia, MO	wheat genetic stock (10)	300

Table 2, continued

Center and location	Crop†‡	No. items
Agric. Exp. Stn., Davis, CA	tomato genetic stock (10)	±2,000
Agric. Exp. Stn., Presque Isle, ME	potato tuber stock (6)	500
Agric. Exp. Stn., Meridian, MS	sorghum, sweet (3)	3,500
Agric. Exp. Stn., Geneva, NY	beans (2)	2,800
	peas (1)	5,000
Agric. Exp. Stn., Madison, WI	carrots (1)	±600
Agric. Res. Center, USDA, PPI, Beltsville, MD	sugarcane (5)	800
Sugarcane Field Stn., USDA, Canal Pt., FL	sugarcane (2)	800
Agric. Exp. Stn., St. Paul, MN	flax (1)	4,000
Delta Br. Exp. Stn., Stoneville, MS	cotton (1)	1,168
USDA Cotton Res. Ctr., Phoenix, AZ	cotton (1)	278

* In all, there are over 100 major laboratory centers where germplasm is maintained by curators and over 150 secondary laboratories as listed by Q. Jones, staff scientist, NPS, ARS, USDA (personal communication 14 Feb. 1975).

† Numbers in parenthesis indicate number of separate species represented. In most instances ARS, CSRS, and AES cooperation is involved. The sorghum collection in Texas is in cooperation with ICRISAT.

‡ The botanical names for the crops are listed in alphabetical order as follows:

alfalfa (*Medicago* spp.)
barley (*Hordeum* spp.)
beans (*Phaseolus* spp.)
bromegrass (*Bromus* spp.)
cabbage (*Brassica oleracea*)
cantaloupe (*Cucumis melo* spp.)
carrots (*Caucus carota* L.)
clover (*Trifolium* spp.)
corn (*Zea mays* L.)
cotton (*Gossypium* spp.)
cowpeas (*Vigna* spp.)
cucumber (*Cucumis sativus* L.)
fescue (*Festuca* spp.)
flax (*Linum* spp.)
lentils (*Lens* spp.)
lettuce (*Lactuca sativa* L.)
millet (*Setaria* spp.)
oats (*Avena* spp.)
onion (*Allium* spp.)
orchardgrass (*Dactylis* spp.)
pea (*Pisum sativum*)
peanuts (*Arachis* spp.)
peppers (*Capsicum* spp.)
potato (*Solanum* spp.)
rice (*Oryza* spp.)
rye (*Secale* spp.)
ryegrass (*Lolium* spp.)
safflower (*Carthanus* spp.)
sorghum (*Sorghum* spp.)
soybean (*Glycine max* L. Merr)
sugar beet (*Beta vulgaris*)
sugarcane (*Saccharium* spp.)
sunflower (*Helianthus* spp.)
sweet clover (*Melilotus* spp.)
tobacco (*Nicotiana tabacum*)
tomato (*Lycopersicon* spp.)
triticale (*X. Triticosecale* spp.)
wheat (*Triticum* spp.)
winter squash (*Cucurbita* spp.)
yams (*Dioscorea* spp.)

regenerate what nature has already provided (Fig. 1). Micke (1975) reports 152 new cultivars have come from mutation breeding.

It is unfair to the many centers in the world where germplasm is now stored to continue to ignore the contribution they have made (Table 1). Vavilov was a world collector and gave Russia a great legacy of germplasm. Likewise, the USDA has for 100 years spent millions of dollars on plant exploration and germplasm preservation as a result of which the flora of the U. S. has been greatly enriched. The Japanese have explored many parts of the world for useful genetic and taxonomic materials and have shared these stocks with many research workers. Germany, Italy, Mexico, Great Britain, and on and on until one includes most of the countries of the world, have been in this business. What has been lacking, and is still lacking, is a means of coordinating the efforts.

You can see that I feel that the prophets of disaster (Harlan, 1972) serve a useful purpose in goading the scientific community, governments, and the general public to a higher level of activity in germplasm conservation. I agree with Harlan when he said, "We are taking risks we need not and should not take" (Harlan, 1972).

Conversely, little or no evidence exists that any of our major crops are threatened with extinction, as these same doom-sayers would have us believe. Reduced productivity locally, or productivity that is economically unacceptable or unsuited to the current social and technological circumstances, yes. But extinction, no. Quinby (1974) regards as a myth the notion that a loss of diversity of vital growth genes is associated with uniformity in sorghum (*Sorghum bicolor* L. Moench) improvement. He says "vital dominant genes are not lost during the plant breeding process—otherwise the plants emerging would be abnormal." In the case of maize, a resurvey after 25 years for races of maize being grown in Central America showed no loss of any race (Frankel, 1973b, p. 76–115). Also, progress continues to be made in productivity in the corn breeding programs of the U. S. (Russell, 1974). Again, only because there was a rich deposit of genes for resistance to crown rust in oats (*Avena sativa* L.) were Frey, Browning, and Simons (1973) able to assemble effective multilines in a host management scheme for widespread use in Iowa and nearby areas. I can say from first-hand evidence and with confidence that the small grain and rice (*Oryza sativa*) collections of the USDA are broad in content of diverse types and widely based pest resistance potential. The same is true for our major crops (Table 2). I cannot remain quiet when these collections are dismissed as "puny."

But no collection, however large, escapes criticism if it fails to contain what a breeder needs at any given time. We should continue to strive to meet this goal, even though it is, in the ultimate, impossible. Enlightened evaluation procedures (Frey et al., 1973; McVey & Roelfs, 1975; Moseman & Smith, 1976) can reduce the burden of collections encumbered with duplication while pointing out items to preserve that assure maximum diversity.

SUMMARY

In apposition to the frenzy about depletion of our germplasm must be set the facts about the world's agricultural goals based on today's needs for food, feed, clothing, shelter, and energy. New cultivars are being made available to better serve man's needs in all continents. These quickly replace less productive and less profitable land races. The world goal is to feed all those who survive—well or badly—upon it. And unless suitable productive cultivars can be bred and maintained, the world's needs cannot be met. Farmers, therefore, cannot choose second-rate cultivars to maintain germplasm diversity. They will of economic necessity grow the most profitable cultivars even though the result is that all grow the same thing over a vast area. The May 9, 1975, issue of *Science* was devoted in its entirety to this subject, and germplasm was referred to repeatedly as one factor in meeting world needs.

In their review, Creech and Reitz (1971) left no uncertainty about the value of our heritage and what present forces are doing to native centers of germplasm. Irreplaceable crop germplasm is being lost from our farms at a rate never before experienced. The main thrust of other analyses is essentially the same.

The challenge to crop and agronomic scientists is to learn how best to cope with the risks and vulnerabilities of plant production. Diversity for diversity's sake has no appeal and is at best of haphazard or chance value. What matters is whether through planned diversity we can make our crops more dependable without sacrificing high productivity and a low cost of production. Research studies of synthetic cultivars, multilines, composites, regional deployment of genes for resistance, avoidance of monoculture, and other schemes to enhance crop dependability have been slow to develop, or unimpressive.

New means of combating insects, diseases, and other pests that involve less dependence on chemicals are also needed. In reducing the damage many pests cause, resistant germplasm holds the most promise. We must not let it slip away.

I suppose most agronomists will agree that something should be done beyond what we are now doing. But agreement and "that something" are two different matters. If I were oratorically endowed, you might respond in a wave of emotional enthusiasm to my plea to meet the challenge. Your response is more apt to be like Jeeter, in the movie "Tobacco Road", who would plow and plant that back 40, raise a corn crop, and cure some of his ills—beginning tomorrow. Much of the world has said germplasm will be looked after tomorrow. Well, *tomorrow is now.*

LITERATURE CITED

Agricultural Research Service. 1974. The regional collection of *Gossypium* germplasm. Report by Technical Committee, Registration Research Project S-77, Genetics and Cytology of Cotton II ARS-H-2 USDA, Washington, D. C. 105 pp.

Anonymous. 1973. N. I. Vavilov Institute of Plant Industry. VIR, Herzen Str., 44 Leningrad U.S.S.R. 15 pp.

Burgess, Sam, ed. 1971. The national program for conservation of crop germplasm. University of Georgia Printing Dep., Athens, Georgia. 73 pp.

Creech, J. L., and L. P. Reitz. 1971. Plant germplasm—now and for tomorrow. Adv. Agron. 23:1–49.

Food and Agriculture Organization. 1970. Wheat and wheat relatives. Plant Int. Newsletter 24. FAO, Rome.

Food and Agriculture Organization. 1972. Barley, oats and rye. Plant Gen. Res. Newsletter 28. FAO, Rome.

Food and Agriculture Organization. 1973. Grain legumes. Plant Gen. Res. Newsletter 29. FAO, Rome.

Frankel, O. H. 1973a. A world survey of genetic resources. Plant Gen. Res. Newsletter 30:25-32. FAO, Rome.

Frankel, O. H., ed. 1973b. Survey of crop genetic resources in their centres of diversity. First report AGP:CGR 73/7. FAO-IBP report. FAO, Rome. 164 pp.

Frankel, O. H., and E. Bennett, ed. 1970. Genetic resources in plants. Handbook No. 11. Blackwell Scientific Publications, Oxford. International Biological Program. 554 pp.

Frey, K. J., J. A. Browning, and M. D. Simons. 1973. Management of host resistance genes to control diseases. Z. Pflanzenkr. Pflanzenschutz 80:160-180.

Harlan, J. R. 1972. Genetics of disaster. J. Environ. Qual. 1:212-215.

Harlan, J. R. 1975. Our vanishing genetic resources. Science 188:618-621.

Hodgson, R. E., ed. 1961. Germplasm resources. Pub. 66. Am. Assoc. for Adv. of Sci., Washington, D. C. 381 pp.

Martin, P. S. 1975. The discovery of America. Science 179:969-974.

McVey, D. V., and A. P. Roelfs. 1975. Postulation of genes for stem rust resistance in the entries of the fourth international winter wheat performance nursery. Crop Sci. 15:335-337.

Micke, A. 1975. Induced mutations in plant breeding. Crop Sci. 15:448.

Moseman, J. G., and R. T. Smith. 1976. Genetic basis for barley germplasm conservation. *In* Third International Barley Genetics Symposium Proceedings. Munich, Germany, July 1975 (in press).

National Academy of Science. 1972. Genetic vulnerability of major crops. NAS-NRC Publication, Washington, D. C.

National Academy of Science. 1975. Plant studies in the People's Republic of China: A trip report of the American Plant Studies Delegation. NAS, Washington, D. C.

Quinby, J. R. 1974. Sorghum improvement and the genetics of growth. Texas A&M Univ. Press, College Station, Texas.

Quisenberry, K. S., and L. P. Reitz. 1974. Turkey wheat: The cornerstone of an empire. Agric. Hist. 48(1):98-110.

Reitz, L. P. 1970. New wheats and social progress. Science 169:952-955.

Russell, A. W. 1974. Comparative performance for maize hybrids representing different eras of maize breeding. Am. Seed Trade Assoc. Res. Report No. 29. p. 81-101.

U. S. Department of Agriculture and National Association of State Universities and Land Grant Colleges. 1973. Recommended actions and policies for minimizing the genetic vulnerability of our major crops. USDA, Washington, D. C. 33 pp.

Wright, J. R. 1975. National plant genetics resources board. Federal Register May 5, 1975, p. 19511.